Technology Transfer, Productivity, and Economic Policy

EDWIN MANSFIELD
University of Pennsylvania

ANTHONY ROMEO
University of Connecticut

MARK SCHWARTZ
Exxon Corporation

DAVID TEECE
Stanford University

SAMUEL WAGNER
Temple University

PETER BRACH
Morgan Guaranty Trust Company

TECHNOLOGY TRANSFER,

PRODUCTIVITY,

AND

ECONOMIC POLICY

W·W·NORTON & COMPANY·NEW YORK·LONDON

THE TEXT OF THIS BOOK *is composed in VIP Baskerville, with display type set in Weiss and Perpetua. Composition and manufacturing are by the Maple-Vail Book Manufacturing Group. Book design by Marjorie J. Flock.*

Library of Congress Cataloging in Publication Data
Main entry under title:
Technology transfer, productivity, and economic
  policy.
  Includes index.
  1. Industrial productivity—United States.
2. Technology transfer—Economic aspects—United
States.   3. Technical assistance, American—Costs.
4. Research, Industrial—United States—Costs.
5. United States—Foreign economic relations.
6. United States—Economic policy.   I. Mansfield,
Edwin.
HC110.152T4   1982   338.973   81–18877
                               AACR2

ISBN 0-393-95222-3

W. W. Norton & Company, Inc. 500 Fifth Avenue, New York, N.Y. 10110
W. W. Norton & Company Ltd. 37 Great Russell Street, London WC1B 3NU

1 2 3 4 5 6 7 8 9 0

Contents

PREFACE

This book is the latest in a series of volumes summarizing the results of a continuing set of studies of the economics of technological change that my students and I have been carrying out. Previous books in this series are *The Economics of Technological Change* (Norton, 1968), *Industrial Research and Technological Innovation* (Norton, for the Cowles Foundation for Research in Economics at Yale University, 1968), *Research and Innovation in the Modern Corporation* (Norton, 1971), and *The Production and Application of New Industrial Technology* (Norton, 1977). The present book extends and builds on the results of these previous works, and presents, as well, findings regarding areas untouched by the earlier books. In particular, it contains new findings regarding the rate, channels, and costs of international technology transfer, the kinds of technology transferred overseas, the benefits of such transfer to the recipients, the effects of international technology transfer on U.S. R and D expenditures, the effects of the composition of an industry's or firm's R and D expenditures on its rate of productivity increase, the size and determinants of imitation costs, the characteristics of the nation's engineering labor force, and the nature and adequacy of federal programs in support of civilian technology.

I am very grateful to the National Science Foundation, which supported these studies. I am also indebted to the many firms who took the time and trouble to provide us with the data and expertise without which these projects could not have been carried out. In all, our empirical results are based on data obtained from literally hundreds of firms and research organizations. Preliminary versions of material contained in this book were presented in papers I gave at the American Economic Association (in 1974, 1978, 1980, and 1981); the Fifth World Congress of the International Economic Association in Tokyo; the Edison Centennial Symposium; the Conference on Economic Growth at the Institute for World Economics at the University of Kiel, Germany; Yale University; the Council of Europe in Strasbourg, France; the Chinese Academy of Social Sciences in Peking, China; the Franklin Institute; the American Chemical Society; the Industrial Research Institute; the Rockefeller Foundation's Villa Serbelloni in Bellagio, Italy; the National Science Foundation's Symposium on the Relationship between R and D and the Returns from Technological Innovation; the ITT Key Issues Lecture Series at New York University; the Symposium on R and D Policy held by the American Association for the Advancement of Science; the Conference on Technology Transfer at the International Institute for Management in Berlin, Germany; the Conference on Postwar Changes in the National Economy held by the National Bureau of Economic Research; the Conference on U.S. Competitiveness at Harvard University; and in testimony before congressional committees; as well as in my 1981 Andersen Lectures at the University of Brussels. I am indebted to various discussants of these papers for their comments.

I am also grateful to Richard Levin of Yale University for his comments on the manuscript. Further, I want to thank the *American Economic Review, Quarterly Journal of Economics, Review of Economics and Statistics, Economic Journal, Economica, Science,* and *Management Science* for allowing the reproduction of material that first appeared in their pages. Also, material has been taken from a paper I was commissioned to write by the Joint Economic Committee of Congress, and which was published by the Committee. Much of Chapter 10 appears (with modifications) in my paper in Devendra Sahal (ed.), *The Transfer and Utilization of Technical Knowledge* (Lexington Books, 1981). Sections also are taken from my paper in Martin Feldstein (ed.), *The American Economy in Transition* (Chicago: National Bureau of Economic Research, 1980).

My co-authors all have been graduate students at the University of Pennsylvania who have done their doctoral dissertations, just recently or some time in the last decade or so, under my direction. Chapter 2 is the work of Romeo, Schwartz, and myself, and Chapter 3 was done by Romeo, Wagner, and myself. Chapter 4 is the work of Teece, and Chapter 5 was done by Teece, Romeo, and myself. Chapter 7 is the work of Schwartz, Wagner, and myself, and Chapter 8 was done by Brach and myself. I did Chapters 1, 6, 9, and 10, as well as the conversion and adaptation of the parts into a book. The responsibility for the topics we explored and the approaches we used is, of course, mine.

EDWIN MANSFIELD

Philadelphia
June 1982

Technology Transfer, Productivity, and Economic Policy

1 TECHNOLOGICAL CHANGE, PRODUCTIVITY, AND INTERNATIONAL TECHNOLOGY TRANSFER

1. Introduction

Technology consists of society's pool of knowledge concerning the industrial, agricultural, and medical arts. It is made up of knowledge concerning physical and social phenomena, knowledge regarding the application of basic principles to practical work, and knowledge of the rules of thumb of practitioners and craftsmen. Although the distinction between science and technology is imprecise, it is important. Science is aimed at understanding, whereas technology is aimed at use. Changes in technology often take place as a consequence of inventions that depend on no new scientific principles. Indeed, until the middle of the nineteenth century, there was only a loose connection between science and technology.

The fundamental and widespread effects of technological change are obvious. Technological change has permitted the reduction of working hours, improved working conditions, provided a wide variety of extraordinary new products, increased the flow of old products, and added a great many dimensions to the life of our citizens. At the same time, technological change also has its darker side. Advances in military technology have enabled modern nation-states to cause human destruction on an unprecedented scale, modern technology has contributed to various kinds of air and water pollution, and advances in industrial technology have sometimes resulted in widespread unemployment in particular occupations and communities. Despite the many benefits that society has reaped from technological change, no one would regard it as an unalloyed blessing.

In this introductory chapter, we take up a number of topics, each of which must be discussed if the studies described in subsequent chapters are to be understood properly. To begin with, we describe the processes of technological change, innovation, and diffusion. Then we touch on the role of technological change in economic growth, and the recent slackening of U.S. productivity growth. Turning to America's position vis-à-vis other nations, we discuss this nation's history of technological leadership, the recent reduction in this technology gap, and the process of international technology transfer. Finally, we describe the recent concerns by policy makers regarding U.S. productivity and civilian technology, as well as regarding international technology transfer.

2. The Production of New Technology

During the past twenty years, economists have begun to study systematically the factors underlying the production of new technology. Their results indicate that the rate of technological change in a particular field depends, to a considerable extent, on the amount of resources devoted by firms, by independent inventors, and by government agencies to the advancement of technology in that field. The amount of resources devoted by the government depends on how closely the field in question is related to the defense, public health, and other social needs for which the government assumes major responsibility; on the extent of the external economies generated by the relevant research and development; and on more purely political factors. The amount of resources devoted by industry and independent inventors depends heavily on the profitability of their use. The total amount that a firm spends on research and development is determined in large degree by the expected profitability of the R and D projects under consideration, and the probability of its accepting a particular R and D project depends heavily on the project's expected returns. Case studies of particular inventions, detailed studies of decision making in particular laboratories, and studies of patent statistics all seem to support this view.[1]

From the proposition that the amount invested by private sources in improving technology in a particular area is influenced by the anticipated profitability of the investment, it follows that the rate of techno-

---

1. See National Bureau of Economic Research, *The Rate and Direction of Inventive Activity* (Princeton: Princeton University, 1962); Edwin Mansfield, *Industrial Research and Technological Innovation* (New York: Norton for the Cowles Foundation for Research in Economics at Yale University, 1968); and Jacob Schmookler, *Invention and Economic Growth* (Cambridge: Harvard University, 1966).

logical change in a particular field is influenced by the same kinds of factors that determine the output of any good or service. For one thing, there are demand factors that influence the rewards from particular kinds of technological change. Thus, if a prospective change in technology reduces the cost of a particular product, increases in the demand for the product are likely to increase the returns from bringing about this technological change. Similarly, a growing shortage and a rising price of the inputs saved by the technological change are likely to increase the returns from bringing it about.

In addition, there are supply factors that influence the cost of making various types of technological change.[2] Obviously, whether people try to solve a given problem depends on whether they think it can be solved, and on how much they think it will cost to solve it, as well as on the expected payoff if they are successful. The cost of making science-based technological changes depends on the number of scientists and engineers in relevant fields and on advances in basic science. In addition, the rate of technological change depends on the amount of effort devoted to making modest improvements that lean heavily on practical experience. Although there is a tendency to focus attention on the major, spectacular inventions, it is by no means certain that technological change in many industries is due chiefly to these inventions rather than to a succession of minor improvements. In recent years, Kenneth Arrow and Samuel Hollander, among others, have stressed the importance of minor improvements and of learning by doing.[3]

Due largely to the activities of the National Science Foundation, the postwar period has seen the development of statistics concerning expenditures on research and development. Although these statistics are not without their shortcomings, they are an important step toward a fuller understanding of the factors determining the rate of technological change. Research and development, as defined by the National Science Foundation, include activities of three kinds. First, there is basic research, which is "original investigation for the advancement of scientific knowledge . . . which do[es] not have immediate commercial objectives."[4] For

2. In addition, the rate of technological change in an industry is affected by the quantity of resources devoted by other industries to the improvement of capital goods and other inputs it uses. Also, it is affected by "spillover" from other industries, the industry's market structure, the legal arrangements under which it operates, the attitudes toward technological change of management, workers, and the public, and many other factors. Of course, there is also a large stochastic element; many inventions occur largely by chance.

3. Kenneth Arrow, "The Economic Implications of Learning by Doing," *Review of Economic Studies* 29 (1962): 155–73; and Samuel Hollander, *The Sources of Increased Efficiency* (Cambridge: MIT Press, 1965).

4. National Science Foundation, *Methodology of Statistics on Research and Development* (Washington, D.C.: Government Printing Office, 1959), p. 124.

example, an economist who constructs an econometric model without any particular application in mind is performing basic research. Firms carry out some basic research, but it is a small percentage of their R and D work. Second, there is applied research, which is research that is aimed at a specific practical payoff. For example, a project designed to determine the properties of a new polymer that a chemical firm plans to introduce would be applied research. The dividing line between basic and applied research is unclear at best. The distinction is based on the purpose of the project, applied research being done to promote specific practical and commercial aims, basic research being done to obtain new knowledge for its own sake.

In addition, there is development, which tries to reduce research findings to practice. Major development projects attempt to bring into being entirely new processes and products. Minor development projects attempt to make slight modifications of existing processes and products. Frequently, prototypes must be designed and constructed, or pilot plants must be built. The dividing line between research and development often is hazy. Research is oriented toward the pursuit of new knowledge, whereas development is oriented toward the capacity to produce a particular product. The outcome of research is generally more uncertain than the outcome of development. Nonetheless, in development there often is considerable uncertainty regarding cost, time, and the profitability of the result.[5]

R and D is, of course, a risky business, as evidenced by a series of studies by the RAND Corporation of weapons development. However, there is a vast difference between military R and D and civilian R and D. In civilian R and D, the technical risks—the risks that the technical objectives of the project cannot be met—tend to be quite modest, due largely to the fact that the bulk of the civilian R and D projects are aimed at fairly modest advances in the state of the art. The commercial risks—the risks that the new product will not be profitable in the market—tend to be much larger than the technical risks in civilian R and D. In other words, it is much more likely that an industrial laboratory can solve the technical problems involved in developing a new or improved product or process than that it will be economically worthwhile to solve these problems.[6]

5. For a detailed study of the development process in polymers, see Edwin Mansfield, John Rapoport, Anthony Romeo, Edmund Villani, Samuel Wagner, and Frank Husic, *The Production and Application of New Industrial Technology* (New York: Norton, 1977), Chapters 4–5.

6. Edwin Mansfield, John Rapoport, Jerome Schnee, Samuel Wagner, and Michael Hamburger, *Research and Innovation in the Modern Corporation* (New York: Norton, 1971).

3. Technological Innovation

A technological innovation is defined as the first commercial intro-duction of new technology. Research and development is only a part of the process leading to a successful technological innovation. In recent years, economists have tended to abandon the traditional two-step model of invention and innovation in favor of a richer model of the process that, if successful, leads to the first commercial introduction of a new product or process. The first part of this process takes place in the inter-val between the establishment of technical feasibility and the beginning of commercial development of the new product or process. This time interval may be substantial (although it is shorter now than fifty years ago). For example, it averaged about a decade for important postwar innovations, like numerical control, freeze-dried foods, and integrated circuits. The second part of this process takes place in the time interval between the beginning of commercial development and the first com-mercial application of the new process or product. This time interval contains a number of distinct stages—applied research, preparation of product specification, prototype or pilot plant construction, tooling and construction of manufacturing facilities, and manufacturing and mar-keting start-up. In all, this time interval seems to have averaged about five years for important postwar innovations.[7]

The management of innovation entails a great deal more than the establishment of an R and D laboratory that turns out a lot of good technical output. In many industries, the bulk of the innovations are not based to any significant extent on the firm's R and D. For example, Sum-ner Myers and Don Marquis found that only a minority of the recent innovations in housing, railroads, and computers were based directly on R and D. They also concluded that technological innovations are more often stimulated by perceived production and marketing problems and needs than by technological opportunities. That is, in the bulk of the cases, it appeared that a market-related stimulus triggered work on a successful innovation. However, it is quite likely that the innovations based on technological opportunities—and on R and D—are the more impor-tant ones. And in any event it is very difficult to sort out the role of technology-push factors from those of market-pull factors.[8]

7. Frank Lynn, "An Investigation of the Rate of Development and Diffusion of Tech-nology in Our Modern Industrial Society," *Report of the National Commission on Technology, Automation, and Economic Progress* (Washington, D.C.: Government Printing Office, 1966).

8. Sumner Myers and Don Marquis, *Successful Industrial Innovations* (Washington, D.C.: National Science Foundation, 1969); and National Science Foundation, *Proceedings of a Conference on Technology Transfer and Innovations* (Washington, D.C.: Government Printing Office, 1967). See B. Klein, *Dynamic Economics* (Cambridge: Harvard, 1977).

Many innovations rely on little in the way of science. But in more and more areas of the economy, innovations have come to depend on a strong scientific base. Merely to be able to imitate or adapt what others have developed, it is necessary for a firm in the aircraft, electronics, or chemical industries to have access to high-quality scientists. Of course, this does not mean that leadership in basic science is a prerequisite for leadership in technology. Unless a country is able to exploit its leadership in basic science, this leadership may not have a great effect on its technology. Moreover, it is not at all accurate to assume that innovations, when they are science-based, must rely on or exploit new science. According to a number of studies, industrial innovations generally are based very largely on relatively old science.[9]

A central problem facing a firm that attempts to be innovative is to effect a proper coupling between R and D, on the one hand, and marketing and production, on the other. The great importance of this problem is emphasized by recent work of Christopher Freeman and Edwin Mansfield.[10] Many R and D projects are designed without sufficient understanding of market and production realities. Many marketing and production people are unnecessarily impervious to the good ideas produced by R and D people. To try to reduce these problems, some firms promote frequent contacts between R and D people and people in other parts of the firm, there being considerable evidence that person-to-person contacts are the most effective way of transferring ideas and technology. Firms also try to break down resistance—in the R and D department and elsewhere—to ideas stemming from outside the firm. Even in R and D-intensive industries like chemicals and pharmaceuticals, a large proportion of the innovations are based on inventions made outside the innovating firm.[11]

Because of the difficulties in carrying out radical innovations in the large, established firms, such innovations often are spearheaded by new firms and "invaders" from other industries. Also, firms where problems and tasks are broken down into specialties and where there is a strict vertical chain of command seem to be less likely to be innovative than firms where there is no strictly defined hierarchy, where communication resembles consultation rather than command, and where individuals have

9. C. Sherwin and R. Isenson, *First Interim Report on Project Hindsight* (Washington, D.C.: Office of the Director of Defense Research and Engineering, 1966); and Illinois Institute of Technology Research Institute, *Technology in Retrospect and Critical Events in Science,* report to the National Science Foundation, 1969.

10. Christopher Freeman, "A Study of Success and Failure in Industrial Innovation," in *Science and Technology in Economic Growth,* ed, B. Williams (London: Macmillan, 1973); Mansfield *et al., The Production and Application of New Industrial Technology, op. cit.;* and Mansfield *et al., Research and Innovation in the Modern Corporation, op. cit.*

11. See the last two references in footnote 10.

to perform tasks in the light of their knowledge of the tasks of the whole firm. Some large firms, like du Pont and ICI, have tried to get some of the advantages of the small firm by creating a number of teams in their development department that operate somewhat like small firms.[12]

4. The Diffusion of Innovations

Once a new process or product is introduced, the diffusion process (the process by which the use of an innovation spreads) begins. The diffusion of a new technique is often a slow process. For example, measuring from the date of first commercial application, it generally took more than ten years for all of the major American firms in the bituminous coal, steel, railroad, and brewing industries to begin using a sample of important new techniques. (Among the innovations included in this sample were the shuttle car, trackless mobile loader, and continuous mining machine in the bituminous-coal industry, and the by-product coke oven, continuous annealing, and continuous wide strip mill in the steel industry.) More recently, similar results have been obtained for the chemical and other industries as well. Also, the rate of diffusion varies widely. Sometimes it took decades for firms to install a new technique, but in other cases they imitated the innovator very quickly. To some extent, these differences may reflect a tendency for the diffusion process to go on more rapidly in more recent times than in the past.[13]

According to the available evidence, the rate of diffusion of an innovation depends on the average profitability of the innovation, the variation among firms in the profitability of the innovation, the size of the investment required to introduce the innovation, the number of firms in the industry, their average size, the inequality in their sizes, and the amount that they spend on research and development. Using these variables, it is possible to explain a large proportion of the variation among innovations in the rate of diffusion. Moreover, this seems to be the case in a wide variety of industries and in other countries as well as in the United States. Econometric models using these variables seem to be useful devices for technological forecasting.[14]

Based on studies of a number of industries, it is clear that firms where the expected returns from the innovation are highest tend to be quickest

12. For discussions of the difficulties in carrying out radical innovations in big firms, see Donald Schon, *Technology and Change* (New York: Delacorte, 1967); and Keith Pavitt, *The Conditions for Success in Technological Innovation* (Paris: Organization for Economic Cooperation and Development, 1971).

13. Mansfield *et al.*, *The Production and Application of New Industrial Technology, op. cit.*

14. *Ibid.*

to introduce an innovation. Also, holding constant the profitability of the innovation, big firms tend to introduce an innovation before small firms. In some industries, this may be due to the fact that larger firms— although not necessarily the largest ones—are more progressive than small firms. But even if the larger firms were not more progressive, one would expect them to be quicker, on the average, to begin using a new technique for reasons discussed in our earlier studies.[15] Also, holding other factors constant, firms with younger and better-educated managers tend to be quicker to introduce new techniques—or at least, this seems to be the case in industries where firms are small.

Firms also differ greatly with regard to the intrafirm rate of diffusion—the rate at which, once it has begun to use the new technique, a firm substitutes it for older methods. A considerable amount of this variation can be explained by differences among firms in the profitability of the innovation, the size of the firm, and its liquidity. Also, there is a tendency for late starters to "catch up." That is, firms that are slow to begin using an innovation tend to substitute it for older techniques more rapidly than those that are quick to begin using it. It is also relevant to note that the same sort of process occurs on the international scene: countries that are slow to begin using an innovation tend to substitute it for older techniques more rapidly than countries that are quick to begin using it. The reasons for this tendency, both at the firm and national levels, seem clear enough.[16]

Sociologists have studied the nature and sources of information obtained by managers concerning new techniques. Judging from the available evidence, firms, once they hear of the existence of an innovation, may wait a considerable period of time before beginning to use it. In many cases, this is quiet rational. But to some extent this may also be due to incomplete or erroneous information, prejudice, and resistance to change. The sources of information sometimes vary depending on how close the manager is to adopting the innovation. For example, in agriculture, mass media are most important sources at the very early stages of a manager's awareness of the innovation, but friends and neighbors are most important sources when a manager is ready to try the innovation. Also, there is evidence of a "two-step flow of communication." The early users of an innovation tend to rely on sources of information beyond their peer group's experience; after they have begun

15. Mansfield, *Industrial Research and Technological Innovation, op. cit.*

16. Edwin Mansfield, *The Economics of Technological Change* (New York: Norton, 1968); G. Ray, "The Diffusion of New Technology," *National Institute Economic Review* 48 (1969):40–83; and L. Nabseth and G. Ray, *The Diffusion of New Industrial Processes* (London: Cambridge University, 1974).

using the innovation, they become a model for their less expert peers, who can imitate their performance.[17]

In addition, a number of other factors also influence the rate of diffusion of an innovation. For example, the diffusion process may be slowed by bottlenecks in the production of the innovation—as in the case of the Boeing 707. Also, the extent of advertising and other promotional activities used by producers of the new product or equipment will have an effect. So too will the innovation's requirements with respect to knowledge and coordination, the diffusion process being impeded if the innovation requires new kinds of knowledge on the part of the user, new types of behavior, and the coordinated efforts of a number of organizations. If an innovation requires few changes in sociocultural values and behavior patterns, it is likely to spread more rapidly. Also, the policies adopted by relevant labor unions influence the rate of diffusion. For example, some locals of the painters' union have refused to use the spray gun. There is, of course, a considerable literature on the effect of collective bargaining on the rate of adoption of new techniques.[18]

5. The Role of Technological Change in Economic Growth

Technological change consists of advances in knowledge concerning the industrial, agricultural, and medical arts. Such advances result in new and improved processes and products, as well as new techniques of organization and management. The fact that technological change plays an important role in permitting and stimulating economic growth seems self-evident. But when one wants to go beyond such bland generalizations to a quantitative summary of the contribution of technological change to the rate of economic growth, a number of basic difficulties are encountered. For one thing, it is hard to separate the effects on economic growth of technological change from those of investment in physical capital, since, to be used, new technology frequently must be embodied in physical capital—new machines and plant. For example, a numerically controlled machine tool (or control mechanism) must be built to take full advantage of some of the advances in the technology related to machine tools. Nor can the effects of technological change easily be separated from those of education, since the social returns from increased

17. Everett Rogers, *Diffusion of Innovations* (New York: Free Press, 1962); J. Bohlen *et al., Adopters of New Farm Ideas,* North Central Regional Extension Publications, No. 13; and E. Katz, "The Social Itinerary of Technical Change," *Human Organization* (1961).

18. See Mansfield, *The Economics of Technological Change, op. cit.*

education are enhanced by technological change, and the rate of technological change is influenced by the extent and nature of society's investment in education.

Despite these and other problems, economists have tried to obtain quantitative measures of the importance of technological change in American economic growth. In a seminal article published in 1957, Robert Solow attempted to estimate the rate of technological change in the nonfarm U.S. economy from 1909 to 1949.[19] His findings indicated that, for the period as a whole, the average rate of technological change was about 1.5 percent per year. Based on these findings, he concluded that about 90 percent of the increase in output per capita during that period was attributable to technological change, whereas only a minor percentage of the increase was due to increases in the amount of capital employed per worker.

Solow's measure of the effects of technological change also included the effects of whatever inputs were excluded, such as increases in education or improved health and nutrition of workers, as well as economies of scale, improved allocation of resources, and changes in product mix. To obtain a purer measure, Edward Denison[20] attempted to include many factors—for example, changes in labor quality associated with increases in schooling—that had been omitted, largely or completely, by Solow and others. Since Denison's study was relatively comprehensive, it resulted in a lower residual increase in output unexplained by the inputs he included than did Solow. Specifically, Denison concluded that the "advance of knowledge"—his term for the residual—was responsible for about 40 percent of the total increase in national income per person employed during 1929–57 in the United States.

6. The Slackening of U.S. Productivity Growth

Policy makers in both the public and private sectors have expressed concern over the reduction in recent years of the rate of productivity growth in the United States. One important productivity measure is output per man-hour. According to the Bureau of Labor Statistics, output per man-hour (in the private sector) grew at an average annual rate of 3.2 percent from 1947 to 1965, but at an average annual rate of only 1.8 percent from 1965 to 1978.

Another important measure of productivity is total factor productiv-

19. Robert Solow, "Technical Change and the Aggregate Production Function," *Review of Economics and Statistics* 39 (1957):312.

20. Edward Denison, *The Sources of Economic Growth in the United States* (New York: Committee for Economic Development, 1962).

ity, defined as real output per unit of total factor input. Since both labor and nonhuman factor inputs are unadjusted for quality changes, the rate of increase of total factor productivity reflects changes in the average quality of resource inputs, as well as changes in technology, changes in efficiency, and other changes discussed in Chapter 6. Total factor productivity increased by 2.5 percent per year between 1948 and 1966, by 1.1 percent per year between 1966 and 1969, and by 2.1 percent per year between 1969 and 1973. Total factor productivity in manufacturing fell sharply from 1973 to 1974, and in 1976 was not much above its 1973 value.[21]

In Chapter 6, we shall look much more closely at the extent of the recent productivity slowdown in the United States. We shall also look carefully at various possible reasons for this slowdown. For now, it is sufficient to point out that some observers believe that this productivity slowdown is due in part to a reduction in the rate of innovation in the United States. In subsequent chapters, we shall describe and evaluate some of the evidence put forth to support this hypothesis.

7. America's History of Technological Leadership

Besides being concerned that the U.S. rate of technological change is slowing down, policy makers are also concerned that America's technological lead over its international competitors is evaporating. Although it is very difficult to measure international differences in technological levels, the available evidence suggests that the United States long has been a leader in technology. Scattered impressionistic evidence indicates that this was true in many fields before 1850. After 1850, the available quantitative evidence indicates that total factor productivity was higher in the United States than in Europe, that the United States had a strong export position in technically progressive industries, and that Europeans tended to imitate American techniques. The existence of such a gap in the nineteenth century would not be surprising, since this was a heyday of American invention. (Among the key American inventions of the period was the system of interchangeable parts.) Needless to say, the United States did not lead in all fields, but it appears that we held a technological lead in many important parts of manufacturing.

21. See Ronald Kutscher, Jerome Mark, and John Norsworthy, "The Productivity Slowdown and the Outlook to 1985," *Monthly Labor Review* 100 (1977):3–8; John Kendrick, "Productivity Trends and Prospects," in *U.S. Economic Growth from 1976 to 1986* (Washington, D.C.: Joint Economic Committee of Congress, October 1, 1976); and Edwin Mansfield, "Technology and Productivity in the United States: Developments and Changes in the Postwar Period," in *The American Economy in Transition,* ed. Martin Feldstein (Chicago: National Bureau of Economic Research, 1980).

After World War II, there was a widespread feeling that this technological gap widened, due in part to the wartime devastation of many countries in Europe and elsewhere. In the 1960s, Europeans expressed considerable concern over the technology gap. They asserted that superior know-how stemming from scientific and technical achievements in the United States had allowed American companies to obtain large shares of European markets in fields like aircraft, space equipment, computers, and other electronic products. In 1966, Italy's foreign minister Amintore Fanfani went so far as to call for a "technological Marshall Plan" to speed the flow of American technology across the Atlantic. In response to this concern, the OECD made a large study of the nature and causes of the technology gap.[22] It concluded that a large gap existed in computers and some electronic components, but that no general or fundamental gap existed in pharmaceuticals, bulk plastics, iron and steel, machine tools (other than numerically controlled machine tools), nonferrous metals (other than tantalum and titanium), and scientific instruments (other than electronic test and measuring instruments). Thus, the OECD studies indicated that the American technological lead was greatest in relatively research-intensive sectors of the economy.

The factors responsible for these technological gaps were difficult to sort out and measure. A host of factors—the social climate, the educational system, the scientific community, the amount and quality of industrial research, the nature of domestic markets, the quality of management, and government policies, among others—influence a country's technological position. According to the OECD studies, the size and homogeneity of the American market was an important factor, but not a decisive one. Also, the large size of American firms was another factor, but not a decisive one. In addition, the large government expenditures on R and D in the United States played an important role. Also, according to the OECD studies, a very important factor was that American firms had a significant lead in the techniques of management, including the management of R and D and the coupling of R and D with marketing and production.

8. The Reduction of the Technology Gap

As readers of publications as diverse as *Business Week*, *Science*, and *Time* are aware, the U.S. technological lead has been reduced in many areas during the past fifteen or twenty years. In some areas, this lead no longer exists at all. Or at least this seems to be the judgment of leading

22. Organization for Economic Cooperation and Development, *Gaps in Technology: General Report* (Paris: Organization for Economic Cooperation and Development, 1968).

engineers, scientists, and managers, both here and abroad. Unfortunately, as indicated in the previous section, it is extremely difficult to measure international gaps in technology. But at least two types of evidence seem to be quite consistent with this view.

First, there is the fact that labor productivity has increased much more slowly in the United States than in western Europe or Japan. Between 1960 and 1977, labor productivity increased by 151 percent in France, 150 percent in West Germany, 64 percent in the United Kingdom, 95 percent in Canada, 279 percent in Japan, and 60 percent in the United States. To some extent, our relatively poor rate of productivity increase has been due to a relatively low rate of investment in plant and equipment. But this is only part of the story. Total factor productivity (which, as we have seen, takes account of both labor and capital inputs) increased more slowly during 1960–73 in the United States than in Canada, France, West Germany, Italy, Japan, Korea, the Netherlands, or the United Kingdom. Although it is difficult to determine how much of this is due to a narrowing of our technological lead, it is reasonable to believe that it is one of the reasons.[23]

Second, the National Science Board has published the results of a study which indicates that the United States originated about 80 percent of the major innovations in 1953–58, about 67 percent of the major innovations in 1959–64, and about 57 percent of the major innovations in 1965–73.[24] Without more information concerning the way in which the sample of innovations was drawn, it is hard to tell whether the apparent reduction in the proportion of industrial innovations stemming from the United States is due to the sampling procedures. Also, without some weighting of the innovations, these data are difficult to interpret. But taken at face value, the findings suggest a reduction in America's technological lead.

In particular areas, there is considerable evidence that foreign firms have cut the U.S. lead substantially. For example, in electronics, the Japanese have become increasingly competitive with U.S. semiconductor manufacturers. According to some observers, Japanese quality standards and productivity now are sometimes more than a match for U.S. manufacturers, even in areas where the United States has been preeminent for decades. In memory chips, the Japanese won a major victory in the market for the 64K RAM, where they had 70 to 80 percent of the market in 1982.

23. L. Christensen, D. Cummings, and D. Jorgensen, "An International Comparison of Growth of Productivity, 1947–73," in *New Developments in Productivity Measurement and Analysis* ed. J. Kendrick and B. Vaccara (Chicago: National Bureau of Economic Research, 1980); and *Science Indicators, 1978* (Washington, D.C.: National Science Foundation, 1979).

24. *Science Indicators, 1974* (Washington, D.C.: Government Printing Office, 1975).

9. International Technology Transfer

Because of the differences among nations in technological levels and capabilities, there is a continual process of international diffusion of technology. Knowledge can be transmitted in various ways–by emigration of engineers and skilled workers, by exports of goods and services, by licensing, and by direct investment, among others. One partial indicator of the amount of technology transferred by the United States, as well as the directions and destinations of the technology flows, is the set of data concerning U.S. receipts and payments of licensing fees and royalties published by the National Science Foundation. In 1977, according to these figures, the United States received about $1.0 billion from unaffiliated foreign residents and about $3.8 billion from foreign affiliates. About two-thirds of these payments came from western Europe and Canada. In 1977, the United States paid about $0.2 billion to unaffiliated foreign residents and about $0.3 billion to foreign affiliates. Based on these figures, American receipts and payments of this sort to unaffiliated foreign residents rose more than $0.6 billion during 1966–77, and those to foreign affiliates rose more than $2.6 billion during the same period.[25]

These figures, although interesting, suffer from many drawbacks. For one thing, payments of this sort between affiliated firms can be influenced by tax and other considerations.[26] For another, a considerable amount of technology is transferred internationally without payment (because the technology is not patented and for other reasons cited in the next section). Nonetheless, at least three important conclusions can be drawn from these data. First, the international transfer of technology is a big and rapidly growing business. Second, multinational firms play a very important role in the international transfer of technology. Third, the fact that our receipts from royalties and licenses far exceed our payments indicates that, although our technological lead seems to have been reduced in many industries, U.S. technology continues to be generally strong.

The data concerning fees and royalties are only one indicator of international flows of technology. Case studies of the diffusion of particular innovations are very useful in this regard. A number of such studies have been carried out in recent years. In general, the results seem to bear out the impression that multinational firms have played a major

25. *Science Indicators, 1978, op. cit.*

26. Also, film-rental receipts and other extraneous (for present purposes) items are included in the U.S. Department of Commerce's data on fees and royalties. See "U.S. International Transactions in Fees and Royalties: Their Relationship to the Transfer of Technology," *Survey of Current Business*, December 1973, pp. 14–18.

role in the transfer process in many industries. For example, John Tilton's study of the diffusion of semiconductors shows that, of the major innovations in semiconductors in the 1960s, American subsidiaries were the first to produce about one-third of them in Britain and about one-fifth of them in France.[27] In particular, subsidiaries of American (and other foreign) firms were relatively quick to use silicon and planar techniques, and, by competing strongly with established European firms, they were able to capture large shares of the market. For example, in 1968, foreign subsidiaries had 44 percent of the market for semiconductors in Britain, 33 percent in France, and 22 percent in Germany. As described by the OECD study of electronic components, "the old patterns of moderate competition have been disrupted in a relatively short time by the development of American direct investment. The change, which occurred around 1962, has been continuing at a rapid pace since and is closely linked with the technological revolutions which took place in the industry, and the failure of many European companies to make the transition to the new technologies early enough."[28]

Multinational firms also transfer technology to the developing countries. For example, Jack Baranson has described the transfer of diesel technology to India.[29] It is important to recognize, however, that these firms face much more difficult problems in transmitting technology to developing countries than to industrialized countries. Many of the techniques of the multinational firms may not be suited very well to the less developed countries, with their plentiful unskilled labor, relatively few skilled labor, and little capital. Moreover, there may be little incentive for multinational firms to adapt their products, production techniques, and marketing methods to the conditions present in developing economies, and (unfortunately) developing economies lack the technical capability to effect the necessary adaptations themselves. The technological gap is so wide that multinational firms find it exceedingly difficult to transfer many technologies to developing countries; and when they manage to effect a technological transplant, its effects often are restricted to narrow segments of the local economy.[30]

27. John Tilton, *International Diffusion of Technology: The Case of Semiconductors* (Washington, D.C.: Brookings Institution, 1971).

28. Organization for Economic Cooperation and Development, *Gaps in Technology: Electronic Components* (Paris: Organization for Economic Cooperation and Development, 1968), p. 115.

29. Jack Baranson, "Transfer of Technical Knowledge by International Corporations to Developing Economies," *American Economic Review* 55 (1966):260–66; and his *Manufacturing Problems in India: The Cummins Diesel Experience* (Syracuse: Syracuse University, 1967).

30. See Keith Pavitt, "Multinational Enterprise and the Transfer of Technology," in *The Multinational Enterprise,* ed. John Dunning (London: Allen and Unwin, 1970); and Christopher Freeman's comments on his paper.

Needless to say, it should not be assumed that the multinational firm is always an American firm taking American technology abroad. For example, when Olivetti took over Underwood, it infused new concepts and technologies into Olivetti Underwood. In recent years, with the trend toward increased foreign direct investment in the United States, some observers have worried that such investment might result in foreign firms gaining access to important American technology. According to the National Academy of Engineering, access to U.S. technology is less important than many other factors in prompting such investment.

10. Channels of International Technology Transfer

One reason why it is so difficult to measure international technology flows is that technology can be transferred across national borders in a variety of ways. As pointed out above, a considerable amount of technology is transferred without any appreciable payment for it. For example, scientists and engineers exchange information at international meetings, and one country's scientists and engineers read the publications of other countries' scientists and engineers. The total flow of technology for which there is no payment is certainly large, but there is presently no way to measure it at all precisely.

Another important channel of international technology transfer is the export of goods. The mere existence or availability of a good in a foreign country may result in the transfer of technology since the good may provide information to the importers of the good. Thus, the export of advanced computers to a particular country may result in technology transfer. In addition, the country may gain technology because the exporters of the good will help the country to use the good efficiently. For example, it may help train workers and others—and this training is a form of technology transfer. Also, if the country that imports the good is able to reverse-engineer it (that is, take it apart to discover how it is constructed), there is, of course, the opportunity for further technology transfer.

As pointed out in the previous section, still another way of transferring technology is through the establishment and utilization of overseas subsidiaries. In a great many cases, firms have become multinational because they have wanted to exploit a technological lead. Once the foreign market was big enough to accommodate a plant of minimum economic size, this decision did not conflict with scale economies. Freight costs and tariffs often hastened such a decision. In some cases, the only way that a firm could introduce its innovation into a foreign market was through the establishment of overseas production facilities. In establishing and utilizing foreign subsidiaries, multinational firms transfer technology in a variety of ways. They train operatives and managers, communicate information and capabilities to engineers and technicians, help the users of their products to use them more effectively, and help suppliers to upgrade their technology.

A firm with a significant new product or process also may engage in licensing agreements with foreigners covering patents, trademarks, franchise, technical assistance, and so on. Licensing agreements often call for the licensee to pay a certain percentage of its sales to the licensor, plus, in some cases, a flat fee for technical help. Some licensing agreements also require the licensee to buy certain inputs from the licensor. Still another way in which technology can be transferred is through the formation of a joint venture, an operation owned jointly by the firm with the technology and a firm or agency of the host country. Joint venture agreements are often made by smaller firms that need capital to complement their technology.[31]

Which of these means of transferring its technology does the innovating firm prefer? According to the available evidence, firms seem to prefer direct investment if they can obtain the necessary resources, and if they believe that licensing will give away valuable know-how to foreign producers who are likely to become competitors in the future. Of course, the longer the estimated life of the innovation, the less inclined a firm is to enter into a licensing agreement. Also, firms prefer direct investment over licensing when the technology is sophisticated and foreigners lack the know-how to assimilate it or when a firm is concerned about protecting quality standards. On the other hand, licensing is often preferred when the foreign market is too small to warrant direct investment, when the firm lacks the resources required for direct investment, or when advantages accrue through cross licensing. Also, in some countries like Japan, direct investment has been discouraged by the government. As for joint ventures, they have advantages with respect to forging good relations with host countries, but they have disadvantages and problems in operation, personnel matters, and the division of profits.[32]

To the host country's government, the choice among these alternative means of obtaining technology looks quite different than it does to

31. See John Dunning, "Technology, United States Investment, and European Economic Growth," in *The International Corporation*, ed. C. Kindleberger (Cambridge: MIT Press, 1970); and Jack Behrman, *National Interests and the Multinational Enterprise* (Englewood Cliffs, N.J.: Prentice-Hall, 1970).

32. See Jack Baranson, "Technology Transfer through the International Firm," *American Economic Review* 60 (1970):435–40; and Louis Wells, "Vehicles for the International Transfer of Technology," in *Technology and Economic Development* (Istanbul: The Economic Research Foundation, 1969).

the firm with the technology. To the host government, direct investment creates many problems because the wholly owned subsidiary of a foreign firm is partly outside its control. The direct investor is only partly responsive to the host nation's economic policies. The investor can draw on funds and resources outside the host country. Moreover, the investor has a global strategy which may be at odds with the optimal operation of the subsidiary from the viewpoint of the host government. Also, the host government cannot be certain of the effect of direct investment on its balance of payments. Joint ventures may overcome some of these disadvantages of direct investment, but they have the disadvantage that the host country must invest more capital. Licensing arrangements eliminate many of the problems of control, but they have the disadvantage that the foreign firm with the technology has little commitment or incentive to help the licensee with managerial and technical problems.[33]

Based on the available evidence, there is considerable variation among industries in the extent to which the international transfer of technology takes place via direct investment rather than by other means. For example, OECD studies indicate that in computers and advanced electronic components (where nearly all of the transferred technology was of American origin), the technology was transferred to Europe by direct investment in about half of the cases and by licensing in about half of the cases. However, in value terms, direct investment was considerably more important than licensing. On the other hand, in plastics and pharmaceuticals (where much of the transferred technology originated in Europe as well as the United States), the situation has been quite different. In pharmaceuticals, direct investment seems to have been the major means of transferring technology within the OECD countries. In plastics, on the other hand, the principal means of transferring technology has been through licensing agreements and joint ventures. According to Keith Pavitt, these differences among industries seem to be due to differences in the extent of the technological lead enjoyed by innovating firms, the extent of competition in the industry, and the extent of specialization of firms in different product areas.[34]

Economic studies also indicate that, with regard to a particular new product, the relative importance of direct investment as a means of transferring technology seems to depend on the stage of the new product's life cycle. For example, consider the case of petrochemicals. Robert Stobaugh has shown that, when various important petrochemicals were relatively new, direct investment was the dominant form of technology transfer, but that as they became mature, licensing became dominant.

33. *Ibid.*
34. Pavitt, *op. cit.*

One reason for this sort of pattern lies in the changes over time in the relative bargaining positions of the innovating firm and the country wanting the technology. When the technology is quite new, it is closely held, and countries wanting the technology are under pressure to accept the firm's conditions, which often are a wholly owned subsidiary. But as time goes on, the technology becomes more widely known, and the host country can take advantage of competition among technologically capable firms to obtain joint ventures or sometimes licenses. Eventually, the technology may become available in plants that can be acquired by the host country on a turnkey basis from independent engineering firms.[35]

11. Recent Concerns over U.S. Productivity and Civilian Technology

As pointed out above, American policy makers have been greatly concerned about the slowdown in the rate of productivity increase in the United States and the narrowing of our technological lead over other countries. There are good reasons for this concern. The slowdown in the rate of productivity increase has meant that potential national output—the amount of goods and services that the United States can produce under full employment—has grown less rapidly than otherwise would be the case. Potential national output is of great importance to the American people, since it sets a limit on their standard of living and the extent to which they can achieve their social and personal goals.[36]

The productivity slowdown has also increased the rate of inflation in the United States. Total cost per unit of output equals total cost per hour of labor divided by output per hour of labor. Thus, the rate of increase of total cost per unit of output equals the rate of increase of total cost per hour of labor minus the rate of increase of labor productivity. (By labor productivity we mean output per hour of labor.) If the rate of increase of labor productivity is high, the rate of increase of total cost per unit of output will be much lower than the rate of increase of total cost per hour of labor. But if the rate of increase of labor productivity is low, the rate of increase of total cost per unit of output will be almost as great as the rate of increase of total cost per hour of labor.

As an illustration, suppose that total cost per hour of labor is increas-

35. Robert Stobaugh, "Utilizing Technical Know-How in a Foreign Investment and Licensing Program," in *Proceedings of the Chemical Marketing Research Association* (1970); and his "Where in the World Should We Put that Plant?" *Harvard Business Review* 46 (1969):129–36.

36. For example, see Council of Economic Advisers, *1979 Annual Report* (Washington, D.C.: Government Printing Office, 1979).

ing at 13 percent per year. If labor productivity is increasing at 3 percent per year, total cost per unit of output will increase at 10 percent per year. On the other hand, if labor productivity is not increasing at all, total cost per unit of output will increase at 13 percent per year. Thus, if prices increase at about the same rate as unit costs, the inflation rate will be 3 percentage points lower if labor productivity is increasing at 3 percent per year than if it is not increasing at all.

This point is widely recognized. Many economists have pointed out that, besides its adverse effects on our rate of economic growth, the slowing of our rate of productivity growth in recent years has exacerbated the problem of quelling inflation. Of course, factors other than the slowing of our rate of productivity growth have been major culprits responsible for the excessive recent rates of inflation in the United States. But this factor nonetheless has been an important one.

Policy makers also are concerned about the reduction in our technological lead over other countries because, in many areas where the United States has a favorable balance of trade, our comparative advantage seems to be based on a technological edge. Needless to say, many factors—such as exchange rates, tariffs, quotas, and the aggressiveness and effectiveness with which firms try to market their products abroad— influence our trade position. But there seems to be widespread agreement among economists and others that the role of technology in U.S. foreign trade is important. Thus, the reduction in our technological lead has meant—and is likely to mean in the future—that some major industries will have more and more difficulty competing in world markets. In the early 1980s, this problem of U.S. competitiveness came increasingly to the fore in Washington and elsewhere.

Faced with the slowdown in productivity and the narrowing of our technological lead, public policy makers have tried to determine what role, if any, the government should play in encouraging or supporting innovation in the private sector. During the late 1970s, the federal government set in motion a Domestic Policy Review on Industrial Innovation, which was a massive study of this question involving participants from government, industry, and the universities. Other studies have been made, both inside and outside the government. A debate has been going on in the executive branch, the Congress, and many parts of the private sector over what public policy toward civilian technology should be. This debate is not new. Its origins can be traced back at least twenty years. In subsequent chapters (particularly Chapter 9), we shall discuss in considerable detail many of the issues in this important area.

12. American Concerns Regarding International Technology Transfer

In the United States, the international transfer of technology, particularly by U.S.-based multinational firms, has become the focus of considerable controversy. Some groups, feeling that the narrowing of our technological lead is due in considerable part to the technological activities of multinational firms, argue that the activities of such firms should be subject to various new kinds of regulations. In their view, multinational firms have been an important factor in transmitting American technology to our foreign competition, thus narrowing our technological lead. Among the most vocal critics of the multinational corporation are the American labor unions. For example, representatives of the AFL-CIO testified in 1971 that "While we share the concern of those who talk about the decline of America's trade balance . . . , we are equally concerned about the export of technology itself."[37] Needless to say, the unions are concerned about the effect of the export of technology on jobs. To them, the multinational firm is a "runaway" corporation, which takes jobs away from the American economy.

More specifically, the AFL-CIO made the following proposals in 1971:

Clear legislative direction is necessary to give the President authority to regulate, supervise, and curb the outflows of U.S. capital. . . . Authority within the President's hands should include consideration for the kind of investment that would be made abroad, the product involved, the country where the investment would be made, the linkage of the investment to the flow of trade and its effect on U.S. employment and the national economy. . . . U.S. government policy has encouraged the export of technology in recent years. . . . This policy should be reversed by giving the President clear authority to regulate, supervise, and curb licensing and patent agreements on the basis of Congressionally determined standards. These would include the kind of investment, the product involved, the country of investment, the linkage to trade flows from such transfers and the effect on U.S. employment and the economy.[38]

Subsequently, legislation was introduced in Congress to create a federal agency to stem the outflow of technology, jobs, capital, and production. Although this legislation was not passed, there has continued to be pressure for actions of this sort. In subsequent chapters, we shall provide data and discussion bearing on this topic.

37. Testimony by A. Biemiller before the Subcommittee on Science, Research, and Development of the U.S. House of Representatives, July 28, 1971.
    38. *Ibid.*

13. Concerns in Host Countries Regarding International Technology Transfer and Multinational Firms

Concern regarding the technological activities of multinational firms has not been limited to the United States. The prevailing attitude toward multinational firms tends to be quite different in the host countries (the countries where such firms have subsidiaries) than in the parent country. To the host countries, the technological activities of multinational firms are often viewed with suspicion and fear. Multinational firms are often viewed as instruments of their parent country's policies. Because of military, political, or economic events, the parent country may induce the multinational firm to withhold its product, reduce its investment in foreign subsidiaries, or alter its policies in other ways that are contrary to the interests of the host country. Also, the host country sometimes fears that the multinational firm, in pursuit of its own profits, may engage in activities that are contrary to the host country's interests and policies. Before granting a license or charter to a firm, host countries generally negotiate with the firm about taxes, amounts of capital to be raised locally, repatriation of profits, and other financial matters. In addition, they sometimes offer incentives for firms to establish R and D laboratories in the country, and apply pressure to get the firms to set up fully integrated plants to upgrade local skills. Also, they often force firms to hire and train local workers, and encourage firms to purchase components from local suppliers.[39]

In western Europe (and Canada), many observers have expressed serious concern in recent decades regarding the technological activities of the multinational firms. They have been worried about a number of aspects of these activities. First, during the 1960s, they expressed considerable concern over American firms taking over European firms, since they feared that, in cases where the European firm was about to make an important innovation, the benefits of the innovation would accrue to the United States rather than Europe. Whether or not this would have been the case depends, of course, on the amount that the American firm paid for the European firm. John Dunning described the conditions under which one might expect that the price would not reflect the firm's true social worth. As he pointed out, many of the reasons commonly given for expecting the price to be below the firm's true worth were fallacious.[40]

39. J. Brian Quinn, "Technology Transfer by Multinational Companies," *Harvard Business Review* 47 (1969):147–61.

40. John Dunning, *Studies in International Investment* (London: Allen and Unwin, 1970).

Second, during the 1960s, they feared that American firms, once they took over European firms, would reduce or eliminate existing R and D efforts carried out in Europe. These activities would be transferred to the firm's central headquarters in the United States. It is difficult to tell whether there was much basis for this concern. American foreign subsidiaries seemed to spend a much smaller percentage of sales on R and D than did their parents in the United States. But according to Dunning, American subsidiaries in Britain spent more on R and D, as a percent of sales, than their British competitors.[41] Moreover, even if it were true, its implications would not have been obvious. As Dunning and others have pointed out, it is not obvious that European countries would have been hurt appreciably if European engineers and scientists were to work for European firms rather than American subsidiaries—particularly if, as some have claimed, there were a shortage of qualified scientists and engineers in some European countries.[42]

Third, they feared that they would become technologically dependent on the United States. For example, the president of the Canadian National Research Council said that, because many Canadian enterprises are "branch plants" and "research is normally done by the parent organization" outside Canada, "Canadian industry has been largely dependent on research in the United States and in Britain. The result of this is that, by comparison with the United States or Britain, relatively little industrial research has been done in Canada by industrial organization."[43] However, it is by no means obvious that direct investment really results in a decrease in the host country's inventive and innovative efforts. Moreover, in technology as elsewhere, there are a great many advantages in international specialization and trade, particularly as the cost of product development in certain areas increases. For example, Pavitt and Quinn have argued that such specialization should form an important part of an enlightened public policy toward science and technology.[44]

Fourth, in contrast to the prevalent view in Canada and many parts of Europe, some countries have not wanted foreign-owned firms to conduct research and development on their soil. For example, Japan has not favored the creation of R and D laboratories by multinational firms. Apparently, one reason for this attitude has been that foreign-controlled

41. John Dunning, "U.S. Subsidiaries and Their U.K. Competitors," *Business Ratios* (1966). Also see D. T. Brash, *American Investments in Australian Industry* (Cambridge: Harvard University, 1966); and A. E. Safarian, *Foreign Ownership of Canadian Industry* (Toronto: McGraw-Hill, 1966).

42. Merton J. Peck, "Science and Technology," in *Britain's Economic Prospects,* ed. R. Caves *et al.* (Washington, D.C.: Brookings Institution, 1968).

43. *Canadian Research Expenditure,* submitted to the Royal Commission on Canada's Economic Prospects, Ottawa, March 8, 1956.

44. Pavitt, *op. cit.;* and Quinn, *op. cit.*

R and D have been considered a subtler type of "brain drain," since the host country's scientists and engineers are being used by foreigners, even though they do not move abroad. However, one obvious difference is that by staying in the country, they continue to create various externalities that may be important. Of course, an important consideration here—as elsewhere in this section—is control. Host countries do not feel comfortable about foreign control of much of their technological resources and capability.[45]

During the 1970s and early 1980s, European fears of technological domination by U.S. multinational firms decreased considerably. And Japanese and other non-U.S. multinational firms became more important in world markets. Nonetheless, many of the concerns regarding the multinational firms have not subsided. In particular, developed countries like Canada are still worried about becoming technologically dependent on other nations, and many developing countries continue to mistrust the multinational firms. In subsequent chapters, empirical findings will be presented that throw new light on the effects on host countries of the technological activities of U.S.-based multinational firms.[46]

14. Summary

The rate of technological change in a particular area depends, to a considerable extent, on the amount of resources devoted by firms, government agencies, and independent inventors to the advancement of technology in that area. The amount invested by private sources for this purpose is influenced both by demand factors (which influence the rewards from particular kinds of technological change) and supply factors (which influence the cost of making various types of technological change). There seems to be a reasonably close relationship between the amount spent on R and D and various measures of inventive output. R and D is a relatively risky activity, but the commercial risks often are greater than the technical risks in civilian R and D.

Research and development are not the only sources of ideas for

45. See Raymond Vernon, *Sovereignty at Bay* (New York: Basic Books, 1971); and Charles Kindleberger, *American Business Abroad* (New Haven: Yale University, 1969).

46. Parts of this chapter are based on Edwin Mansfield, "Technology and Technological Change," in *Economic Analysis and the Multinational Enterprise*, ed. John Dunning (London: Allen and Unwin, 1974); Edwin Mansfield, "The Competitive Position of U.S. Technology," Conference on U.S. Competitiveness, Harvard University, 1980; Edwin Mansfield, "Economic Growth or Stagnation: The Role of Technology," in *Looking Ahead* (Washington, D.C.: National Planning Association, Spring 1980); and Edwin Mansfield, "Research and Development, Productivity, and Inflation," *Science*, September 5, 1980, 1091–93.

industrial innovations. In many industries, the bulk of the innovations are not based to any significant extent on the firm's R and D. One of the most important problems facing a firm that attempts to be innovative is to effect a proper coupling between R and D, on the one hand, and marketing and production, on the other. The diffusion of a major innovation frequently has taken over a decade in the United States. The rate of diffusion depends on the average profitability of the innovation, the variation among firms in the profitability of the innovation, the size of the investment required to introduce the innovation, the number of firms in the industry, their average size, the inequality in their sizes, the amount they spend on R and D, and other factors.

It is very difficult to separate the effects on economic growth of technological change from those of other factors like investment in physical capital or education. Nonetheless, according to the best available estimates, the advance of knowledge was responsible for about 40 percent of the total increase in national income per person employed during 1929–57 in the United States. In recent years, there has been a marked slowdown in the rate of increase of U.S. productivity. In the private sector, output per man-hour grew at an average annual rate of 3.2 percent from 1947 to 1965, but at an average annual rate of 1.8 percent from 1965 to 1978. Although there is considerable uncertainty concerning the reasons for this slowdown, some observers attribute it partly to a slowdown in the rate of technological change and the rate of innovation in the United States.

Because of the differences among nations in technological levels and capabilities, there is a continual process of international technology transfer. There are many channels of international technology transfer, including the emigration of engineers and skilled workers, the export of goods and services, licensing, direct investment, and others. The prevailing attitude toward technology transfer via multinational firms tends to be quite different in the host countries than in the parent countries. To the host countries, the technological activities of multinational firms are often viewed with suspicion and fear. Many host countries fear that they are (or that they will become) technologically dependent on the parent countries. Also, they are concerned that the multinational firms may be able to siphon off practically all of the benefits of the transferred technology.

In the United States, policy makers, faced with a slowdown in productivity growth and the narrowing of our technological lead over other countries, have tried to determine what role, if any, the government should play in encouraging or supporting innovation in the private sector. A debate has been going on in the executive branch, Congress, and many parts of the private sector over what public policy toward civilian technology should be. A somewhat different, but related, debate also has been going on concerning international technology transfer. Some groups, feeling that the narrowing of America's technological lead is due in considerable part to the transfer of technology by U.S.-based multinational firms, argue that the activities of such firms should be subject to various new kinds of regulation. Others retort that such regulations are unnecessary and unwise.

2 INTERNATIONAL TECHNOLOGY TRANSFER: RATES, LEAKAGES, AND BENEFITS TO RECIPIENTS

1. Introduction

How rapidly do industrial innovations spread from one country to another? What factors determine whether a particular country is relatively quick or slow to begin producing a particular innovation? To understand the process of international technology transfer, it is essential that we have answers to these questions. In the first part of this chapter, we summarize some earlier work on this score and present some new findings that should help to illuminate these questions.

As pointed out in the previous chapter, there has been considerable controversy over the transfer of technology by U.S.-based multinational firms to their overseas subsidiaries.[1] Some groups in the United States, feeling that the apparent narrowing of the American technological lead is due in part to such transfers, argue that the technological activities of such firms should be subjected to various new kinds of regulations. Other groups deny that such measures are necessary or desirable. In the countries where the subsidiaries are located, there is sometimes a feeling that the terms on which the transfers occur are inequitable, and that there are dangers in becoming too dependent on outside technology.

Despite the enormous amount of attention that has been devoted to

1. For discussions of these controversies, see A. Biemiller, *op. cit.* (cited in Chapter 1, footnote 37); Richard Caves, "Effect of International Technology Transfers on the U.S. Economy," in National Science Foundation, *The Effects of International Technology Transfers on U.S. Economy* (Washington, D.C.: Government Printing Office, 1974); and Gary Hufbauer, "Technology Transfers and the American Economy," in National Science Foundation, *The Effects of International Technology Transfers on U.S. Economy* (Washington, D.C.: Government Printing Office, 1974).

this topic, very little is known about the nature of the technology that is being transferred overseas in this way, the extent to which it leaks out to non-U.S. competitors, the size of the benefits it confers on the host (and other non-U.S.) countries, and the sorts of non-U.S. firms that receive the largest benefits of this sort.[2] In the second part of this chapter, we present some findings that shed new light on each of these matters.

2. Technology Transfer: Forms and Problems

At the outset, it is worthwhile to make a number of distinctions. First, *vertical* technology transfer should be distinguished from *horizontal* technology transfer. Vertical technology transfer occurs when information is transmitted from basic research to applied research, from applied research to development, and from development to production. Such transfers occur in both directions, and the form of the information changes as it moves along this dimension. Horizontal technology transfer occurs when technology used in one place, organization, or context is transferred and used in another place, organization, or context. The problems involved in transferring technology from one country to another are quite different when the transfer is vertical as well as horizontal. In general, the difficulties and costs are much greater under these circumstances than if only a horizontal transfer is involved.

Second, it frequently is useful to differentiate between *general* technology (information common to an industry or trade), *system-specific* technology (information concerning the manufacture of a certain item or product that any manufacturer of the item or product would obtain), and *firm-specific* technology (information that is specific to a particular firm's experience and activities, but that cannot be attributed to any specific item the firm produces). Also, it is worthwhile to distinguish between several phases of the process of technology transfer. In the past, the first phase frequently has been the export of a new material or product by one country to another. This phase, sometimes called *material* transfer, has often been followed by *design* transfer, which is the transfer of designs, blueprints, and the ability to manufacture the new material or product in the recipient country. Finally, there is the phase of *capacity* transfer, which occurs when the capacity to adapt the new item to local conditions

2. For an evaluation of the current state of the art in this area, see the articles by Richard Caves, Gary Hufbauer, Keith Pavitt, and Robert Stobaugh in National Science Foundation, *ibid*. Also see Baranson, "International Transfers of Industrial Technology by U.S. Firms and Their Implications for the U.S. Economy," Report to the U.S. Department of Labor, 1976; and his "Technology Transfer through the International Firm" (cited in Chapter 1, footnote 32).

is transferred. Clearly, this last phase—the phase of learning how to learn as well as to use what others have learned—is quite different from the earlier phases and much more difficult and costly to achieve. It is a phase that many countries have yet to enter in the more sophisticated areas of technology.

It is also important to recognize the kinds of problems involved in the transfer of technology. Some people assume that technology is merely a sheaf of blueprints and that all one has to do is ship off the right set of papers. Unfortunately, it isn't that simple. For one thing, the available evidence, both for recent and for earlier periods, indicates that publications and reports are a much less effective way of transferring technology than the movement of people. To transfer "know-how," much of which is not written down in any event, there is frequently no substitute for person-to-person training and assistance, some of which may have to go on for extensive periods of time.

Also, the technology that is transferred must often be adapted, if it is to meet the needs of the recipient. Because markets in the recipient country tend to be smaller, because input prices are not the same, because the vendor infrastructure is different with regard to capability, cost, and quality, because of national differences in taste, climate, and so forth, the technology must be adapted. This process of adaptation can be more difficult than it may seem. Just as in the original innovation, it is very important that a proper mating occurs between the technological considerations on the one hand and the more purely economic considerations on the other.[3]

3. Rates of International Technology Transfer

Very little information is available concerning the rates at which technology is transferred from one country to another. In a relatively early study, Gary Hufbauer[4] provided data concerning the rate at which synthetic materials spread. His interest centered primarily on design transfer rather than material or capacity transfer. Specifically, he provided the date at which a number of synthetic materials were first produced in selected countries. Using these data, he calculated for each country the average imitation lag—that is, the average length of time

3. See George Hall and Robert Johnson, "Transfer of Aerospace Technology to Japan," in *The Technology Factor in International Trade,* ed. Raymond Vernon (New York: National Bureau of Economic Research, 1970); Y. Hayami and Vernon Ruttan, *Agricultural Development and International Perspective* (Baltimore: Johns Hopkins, 1971); and Mansfield *et al., The Production and Application of New Industrial Technology* (cited in Chapter 1, footnote 5).

4. Gary Hufbauer, *Synthetic Materials and the Theory of International Trade* (Cambridge, Mass.: Harvard University, 1966).

that this country lagged behind the innovating country in beginning to produce a new synthetic material. Some of Hufbauer's results are shown in Tables 2.1 and 2.2. As you can see, the countries that are leaders for one innovation (frequently Germany and the United States) tend to be leaders for other innovations as well. This is not surprising, since the basic factors underlying world technological leadership in synthetic materials are likely to persist, at least for reasonably long periods of time.

In 1971, John Tilton[5] carried out a study of the international diffusion of semiconductor technology. Like Hufbauer, he was interested primarily in design transfer. His results, shown in Table 2.3, indicate how quickly France, Germany, Japan, the United Kingdom, and the United States began producing various semiconductor innovations. If one computes the average lag between the date of innovation and the time when each country began producing the innovation, the results indicate that the average lag was 0.1 years for the United States, 2.2 years for Great Britain, 2.5 years for Japan, 2.7 years for Germany, and 2.8 years for France. Since semiconductors originated in the United States, it would be expected that the average U.S. imitation lag would be shorter than that for other countries. Based on the data in Table 2.3, semiconductor technology spread rapidly from one nation to another.

Both among synthetic materials and among semiconductors, there are substantial differences in how rapidly a particular country begins to produce a new product. For example, whereas the United Kingdom was eleven years behind the world innovator in producing cellulose acetate, the United Kingdom led all the others by four years in producing high-pressure polyethylene. And whereas Japan was eight years behind the world innovator in producing the surface barrier transistor, it led all the others by a year in producing the tunnel diode. Thus, although there is some correlation between how rapidly a country begins producing one new product and how soon it produces another, it would be a great mistake to think that a country's position in this regard is fixed and immutable. On the contrary, there is a considerable amount of variation, which is analyzed in the next section.[6]

4. Factors Associated with Imitation Lags

What factors seem to be associated with how rapidly a country begins to produce a new product? The following hypotheses seem reasonable.

5. See Tilton, *op. cit.* (cited in Chapter 1, footnote 27).

6. Besides the studies by Hufbauer and Tilton, a number of studies have been carried out of imitation lags in the pharmaceutical industry. For example, see William Wordell, "Introduction of New Therapeutic Drugs in the United States and Great Britain: An International Comparison," *Clinical Pharmacology and Therapeutics* 14 (1973). However, this industry is affected very greatly by the regulatory policies of various countries.

TABLE 2.1

*Year of First Commercial Production of Selected Major
Synthetic Materials, Selected Countries, 1905–63*

MATERIAL	UNITED STATES	GERMANY	UNITED KINGDOM	FRANCE	ITALY	JAPAN	BELGIUM	SPAIN
Cellulose acetate	1908	1905	1916	1912	1936	1927	1935	1960
Cellophane	1924	1925	1930	1917	1946	1929	1925	1949
Phenol formaldehyde	1909	1910	1910	1916	1922	1923	1934	1941
Urea formaldehyde	1929	1929	1928	1930	1936	1935	1938	1947
Melamine formaldehyde	1939	1935	1938	1955	1951	1951	—	1960
Alkyd	1926	1927	1929	1928	1927	1931	1938	1945
Polyvinyl chloride	1933	1931	1940	1940	1951	1939	1949	1950
Polystyrene	1933	1930	1950	1951	1942	1957	—	1958
High-pressure polyethylene	1941	1944	1937	1954	1952	1954	1962	1963
Linear polyethylene	1956	1955	1959	1956	1954	1958	—	—
Polypropylene	1957	1957	1959	1960	1957	1961	—	—
Polymethyl methacrylate	1936	1930	1933	1938	1937	1938	—	1950
Neoprene	1931	1956	1960	—	—	1962	—	—
Styrene rubber	1941	1953	1958	1956	1957	1960	—	—
Polyurethane rubber	1954	1941	1942	1951	1955	1956	1954	1955
High-tenacity rayon	1937	1935	1936	1936	1939	1941	1949	1941
Nylon filament	1938	1949	1941	1941	1938	1942	1955	1953
Acrylic fiber	1944	1943	1957	1955	1958	1957	1961	1963
Polyester fiber	1949	1955	1950	1954	1954	1958	—	1961

Source: Gary Hufbauer, *Synthetic Materials and the Theory of International Trade* (Cambridge, Mass.: Harvard University, 1966), pp. 131–32. Note that only a sub- set of the materials and countries covered by Hufbauer is included here.

TABLE 2.2

Average Imitation Lags in Synthetic Materials,
Selected Countries, 1910, 1930, and 1950[a]

COUNTRY	1910			1930			1950			
	PLASTICS	FIBERS	TOTAL	PLASTICS	FIBERS	TOTAL	PLASTICS	RUBBERS	FIBERS	TOTAL
United States	0.7	13.7	3.3	3.1	6.2	5.7	2.6	0.2	7.5	4.6
Germany	7.2	0.4	5.8	4.1	0.9	1.5	2.1	9.3	1.1	2.5
United Kingdom	7.0	9.9	7.6	6.1	1.1	1.9	7.7	9.8	5.7	6.9
France	4.6	3.3	4.3	4.5	4.0	4.1	7.4	9.8	9.2	8.6
Italy	36.7	15.7	32.5	28.9	19.4	20.8	15.8	9.8	14.6	14.5
Japan	34.9	15.1	30.9	23.8	16.6	17.7	12.1	9.8	15.6	13.6
Belgium	36.7	13.0	32.0	34.3	22.8	24.5	17.7	9.8	18.9	17.4
Spain	36.7	13.6	32.1	34.7	16.8	19.5	23.9	9.8	28.1	24.8

Source: Hufbauer, op. cit., p. 137.

[a] In each year, the imitation lag for each product is weighted by that product's share of current world exports in that year. The average imitation lag is obtained by summing the results.

TABLE 2.3

Year of First Commercial Production of Major
Semiconductor Devices, by Country, 1951–68

DEVICE	U.S.	U.K.	JAPAN	GERMANY	FRANCE
Point contact transistor	1951	1953	1953	1953	1952
Grown junction transistor	1951	—	1955	1953	1953
Alloy junction transitor	1952	1953	1954	1954	1954
Surface barrier transistor	1954	1958	1962	—	—
Silicon junction transistor	1954	1958	1959	1955	1960
Diffused transistor	1956	1959	1958	1959	1959
Silicon controlled rectifier	1956	1957	1960	1960	1960
Tunnel diode	1958	1960	1957	1960	1960
Planar transistor	1960	1961	1961	1962	1963
Epitaxial transistor	1960	1962	1961	1962	1963
Integrated circuit	1961	1962	1962	1965	1964
MOS transistor	1962	1964	1963	1965	1964
Gunn diode	1963	1965	1965	1967	1965

Source: John Tilton, *International Diffusion of Technology: The Case of Semiconductors* (Washington, D.C.: Brookings Institution, 1971).

First, how rapidly a country begins producing a new product would be expected to be directly related to how much the country spends on research and development in the relevant industry. Based on our earlier studies, there is evidence that firms that spend relatively large amounts on R and D tend to be relatively quick imitators (as well as relatively frequent innovators). The same sort of relationship would be expected to hold true for entire countries. Countries that spend relatively large amounts on R and D in the relevant industry are likely to be close to the technological frontiers and able to imitate quickly, whereas other countries may not be able to do so.

Second, there is likely to be a tendency for imitation rates to be faster for more recent innovations than for earlier ones. This hypothesis has been advanced by Richard Cooper, among others. Because of improvements in communications, transportation, and methods for evaluating investments in facilities to produce new products, one might expect such a trend toward higher international diffusion rates. Third, how rapidly a country begins producing a new product might be expected to be affected by the level of concentration in the relevant industry in this country. In some industries, decreases in concentration might mean more rapid imitation because they might mean more competition. In other industries, decreases in concentration might mean less rapid imitation because they might mean that the relevant industry would be too fragmented to allow any member to amass the necessary resources quickly.[7]

7. To some extent, the line of causation may run in the opposite direction. How rapidly an industry begins to produce new products may influence its level of concentration.

To test these hypotheses, data were obtained concerning thirty-seven innovations in the plastics, semiconductors, and pharmaceutical industries.[8] For each innovation, we found the date of its first production in five major countries—France, Germany, Japan, the United Kingdom, and the United States. (For plastics, Italy was included too.) Based on these data, we calculated the lag between the year when each innovation was first produced in the innovating nation and the year when it was first produced in the jth nation. For the ith innovation, this lag (in years) is designated as l_{ij}. Based on OECD data, we obtained estimates of each of these nations' R and D expenditures in these industries, and calculated R_{ij}, the jth nation's percentage of the total R and D expenditures of these countries (in the industry in which the ith innovation occurred).[9] Estimates were also made of C_{ij}, the four-firm concentration ratio in the jth nation (in the industry in which the ith innovation occurred).[10]

In each industry, we assumed that

$$(2.1) \qquad l_{ij} = \alpha_0 + \alpha_1 R_{ij} + \alpha_2 t_i + \alpha_3 C_{ij} + z_{ij},$$

where t_i is a measure of when in time the ith innovation occurred. Specifically, t_i equals one plus the year when the ith innovation occurred minus the year when the first innovation in the sample occurred. Of course, the α's are parameters, and z_{ij} is a random error term. Using least-squares, we estimated the α's in each industry. The resulting estimates are shown in Table 2.4.[11]

These results have at least three interesting implications. First, in the

---

However, it would take us too far afield to try to deal with this possible identification problem here.

8. These innovations were polyester, silicone, polystyrene, polystyrene/styrene/butadiene, polystyrene/styrene/acrylonitrile, acrylonitrile/butadiene/styrene, high pressure polyethylene, linear polyethylene, polypropylene, methyl methacrylate, nylon, fluoroethylene, acetal, polycarbonate, spandex, point contact transistor, grown junction transistor, alloy junction transistor, surface barrier transistor, silicon junction transistor, diffused transistor, silicon controlled rectifier, tunnel diode, planar transistor, epitaxial transistor, integrated circuit, MOS transistor, Gunn diode, fortral, gentalline, aldomet, dolicin C, indocin, zyloric, septrin, stelazine, and thorazine.

9. The data concerning R and D expenditures pertain to 1963 or 1964. To convert to dollars, we used the findings of A. S. MacDonald, "Exchange Rates for National Expenditure on Research and Development," *Economic Journal* 83 (1973): 477–85. The resulting figures are crude, but no better data seem to exist.

10. For semiconductors, these data came from Tilton, *op. cit.* For plastics, they come from a variety of sources, including Organization for Economic Cooperation and Development, *Gaps in Technology: Plastics* (Paris: Organization for Economic Cooperation and Development, 1968). For pharmaceuticals, they come from a variety of sources, including studies of concentration by the European Economic Community.

11. The data regarding l_{ij} come from Hufbauer, *op. cit.;* Tilton, *op. cit.;* Paul deHaen's publications concerning new pharmaceuticals; and correspondence with the relevant firms.

TABLE 2.4

Least-Squares Estimates of the Parameters in Equation (2.1)[a]

PARAMETER	PLASTICS	SEMICONDUCTORS	PHARMACEUTICALS
α_0	23.21	2.14	−2.31
	(5.85)	(1.26)	(1.21)
α_1	−0.21	−0.35	0.016
	(5.08)	(2.37)	(1.00)
α_2	−0.01[b]	−0.008	−0.047
	(4.73)	(1.34)	(0.67)
α_3	−0.20	0.012	0.139
	(2.69)	(0.64)	(2.39)
n	77	62	41
R^2	0.39	0.38	0.07

[a] The number in parentheses under each estimate is its t-ratio.
[b] In the plastics industry, if t_i^2 (rather than t_i) is used as an independent variable, the fit is improved. The estimate given above is based on the use of t_i^2.

plastics and semiconductor industries, there is a highly significant tendency for countries that spend relatively large amounts on R and D to have relatively short imitation lags. On the average, a 10-percentage-point increase in a country's share of total R and D in the industry is associated with a reduction in the imitation lag of about two or three years. In the pharmaceutical industry, there appears to be no such tendency, perhaps because of regulatory considerations.[12] Second, in all three industries, when other factors are held constant, imitation lags tend to decrease with time. The extent of the decrease seems smallest in semiconductors and greatest in plastics.[13] Third, the effects of industrial concentration on a country's imitation lag seem to be mixed. In pharmaceuticals, increased concentration seems to be associated with longer imitation lags; in plastics, the reverse seems to be true; and in semiconductors, there seems to be no statistically significant relationship between them.[14]

12. For a discussion of regulation in this industry, see Robert Helms, ed., *The International Supply of Medicine* (Washington, D.C.: American Enterprise Institute, 1980).

13. As indicated in note b to Table 2.4, t_i^2 seems to fit better than t_i in plastics. However, this trend cannot continue much longer since there is a limit to how short the lag can be.

14. How rapidly a country begins producing a new product may depend on the channel through which it receives the new technology. Based on correspondence and interviews with firms, as well as on the published literature, we determined the channel of transfer in each case. Letting D_{ij} be a dummy variable that equals 1 if the first producer of the ith innovation in the jth country was created by direct investment of a multinational firm based outside country j, and 0 otherwise, we used D_{ij} as an additional independent variable in equation (2.1). It was not statistically significant in any industry.

5. Age of Technology Transferred Overseas

As pointed out earlier, some observers are concerned that the transfer of technology by U.S.-based firms may increase the likelihood that this technology will fall into the hands of non-U.S. firms that will use it to compete more effectively with U.S.-based firms. To those who feel this way, a particular worry is that many U.S.-based firms may be transferring their most *modern* technology overseas. However, although Jack Baranson has presented some case studies that suggest that U.S.-based firms are becoming more willing than in the past to transfer their newest technology overseas, there is very little published information on this score. In other words, we really do not know whether there has been a significant change in the age of the technology that is transferred. To help fill this important gap, we obtained information concerning the age of the technology transferred in a sample of sixty-five cases.

To obtain these cases, we chose a carefully selected sample of 31 U.S.-based firms. Twenty-six of these firms were chosen essentially at random from the largest 500 manufacturing firms; 2 were chosen from the second largest 500 manufacturing firms, and 3 were chosen from those not in the 1,000 largest manufacturing firms. Once this sample was chosen, one or more of the technologies that each firm had transferred overseas during 1960 to 1978 were selected, also essentially at random.[15] Interviews were carried out with senior executives of each firm, and the firm's internal records were used to determine the year when each technology was first introduced in the United States. The number of years that elapsed between the time when the technology was first introduced in the United States and the year when it was transferred overseas is used as a measure of the age of the technology when it was transferred.

As shown in Table 2.5, the mean age of the technologies transferred to overseas subsidiaries in developed countries was about six years, which was significantly less than the mean age of the technologies transferred to overseas subsidiaries in developing countries (about ten years). This difference is not surprising. Because many newer technologies are inap-

15. The sample covers a wide variety of industries, including chemicals, drugs, petroleum, electrical equipment and electronics, machinery, instruments, glass, food, and rubber. The number of firms in the sample is divided between the largest five hundred firms and the second largest five hundred firms in proportion to the total sales in each group. The number of cases obtained from each firm did not vary greatly, and averaged about two. One technology was transferred overseas before 1960. By essentially at random, we mean that these samples were not chosen on the basis of a table of random numbers, but that they were a systematic sample from a list or that there were so few relevant items that the choice was automatic.

TABLE 2.5

*Mean and Standard Deviation of Number of Years
between Technology's Transfer Overseas and Its Initial Introduction
in the United States, Sixty-five Technologies*

CHANNEL OF TECHNOLOGY TRANSFER	MEAN (YEARS)	STANDARD DEVIATION (YEARS)	NUMBER OF CASES
Overseas subsidiary in developed country	5.8	5.5	27
Overseas subsidiary in developing country	9.8	8.4	12
Licensing or joint venture	13.1	13.4	26

propriate for developing countries or are difficult and expensive to transfer there, we would expect the technology transferred to developing countries to be older than that transferred to developed countries. Table 2.5 also suggests that the mean age of the technologies transferred through licenses, joint ventures, and other channels other than subsidiaries tends to be higher than the mean age of the technologies transferred to subsidiaries. (This is consistent with our finding in Chapter 3 that firms tend to transfer their newest technology overseas through subsidiaries rather than licensing or joint ventures, but that the latter channels become more important as the technology gets older.)[16]

Are the recently transferred technologies newer than those transferred a number of years ago? To find out, we used χ^2 tests to see whether the proportion of transferred technologies that were less than five years old (at the time of transfer) was greater during 1969–78 than during 1960–68.[17] For technologies transferred to subsidiaries in developed countries, this was the case, and the increase in this proportion (from 27 percent in 1960–68 to 75 percent in 1969–78) was both large and statistically significant. But for technologies transferred to subsidiaries in

16. Based on a *t*-test, the mean age of the technologies transferred to subsidiaries in developed countries is significantly smaller (at the .05 significance level) than the mean age of the technologies transferred to subsidiaries in developing countries. The standard deviation of the ages of the technologies transferred through licensing and joint ventures is significantly greater (at the .05 significance level) than that of the ages of the technologies transferred to subsidiaries in developed countries. Thus, the Aspin-Welch or Dixon-Massey test had to be used to test whether the mean age of the technologies transferred to subsidiaries in developed countries is less than the mean age of technologies transferred through licensing and joint ventures. Both tests indicate that this difference is significant at the .05 level, which is used throughout this chapter.

17. More accurately, this test is carried out for technologies transferred to subsidiaries in developed countries. For technologies transferred to subsidiaries in developing countries, and for technologies transferred through licensing and joint ventures, the observed difference in the proportion is not in accord with this hypothesis, so there is no need for such a test.

developing countries or for those transferred through channels other than subsidiaries, there appeared to be no such tendency, at least in this sample. The fact that the technologies licensed to, or jointly exploited with, non-U.S. firms were no newer in 1969–78 than in 1960–68 is worth noting, since some observers worry that U.S. firms may have come to share in this way more and more of their newest technologies with foreigners.[18]

6. Leakage of Technology and Competitive Response by Non-U.S. Firms

As noted above, another major concern of some observers is that the transfer of technology to overseas subsidiaries will hasten the time when non-U.S. producers have access to this technology. Yet no systematic studies have been carried out to determine how rapidly such technology "leaks out" and the extent to which such transfers actually hasten its leaking out. To help fill this void, we obtained estimates of this sort for twenty-six technologies chosen essentially at random from those included in the previous section (which were transferred to subsidiaries).[19] In each case, the U.S.-based firm estimated how soon (after it was transferred) the technology became known to one or more of its non-U.S. competitors. Although it frequently is difficult for firms to make very precise estimates of this sort (since they do not know exactly when their competitors acquire the relevant information), these firms (with all the benefits of hindsight) felt that they could provide reasonably reliable estimates, based on direct knowledge concerning their competitors' information-gathering activities as well as on data regarding the time when their competitors set in motion relevant investment projects and introduced imitative products or processes. For present purposes, these estimates should be quite adequate.

The results, shown in Table 2.6, indicate that the mean lag between the technology transfer and the time when some non-U.S. competitor had access to the technology was about four years.[20] However, because

18. Davidson and Harrigan have presented valuable evidence concerning the age of technologies transferred abroad by U.S.-based firms, at the time when they were first transferred abroad. Their results, like ours, suggest that the average age of technology transferred abroad has declined. Note that our data do not pertain only to the initial transfer abroad of a technology. See W. Davidson and R. Harrigan, "Key Decisions in International Marketing: Introducing New Products Abroad," *Columbia Journal of World Business* 12 (1977):15–23.

19. Fifteen firms are included. The technologies span a wide range of American industry, including chemicals, drugs, petroleum, machinery, food, and rubber. For the way in which the sample was chosen, see footnote 15.

20. One of the simplest models that might be used to represent this leakage process is that the probability of a technology's leaking out between time t and time $t + \Delta$ is pro-

TABLE 2.6

*Number of Years between Technology Transfer
and the Time When the Technology Became Known to
Some Non-U.S. Competitor, Twenty-six Technologies*

NUMBER OF YEARS	NUMBER OF TECHNOLOGY TRANSFERS[a]
Zero and under 0.5	4
0.5 and under 1.5	7
1.5 and under 4.5	5
4.5 and under 6.5	4
6.5 and over	6

[a]In some cases, the technology had not become known to non-U.S. competitors at the time the data were collected; thus, we only have lower bounds on the relevant time interval. In these cases, we assume in this table that they would be divided among the class intervals above this lower bound in proportion to the number of cases in each of the relevant class intervals. This assumption is innocuous and is not made in the analysis of the data. See footnote 22.

non-U.S. firms gained access to many of these technologies relatively soon after they were transferred, one cannot conclude that the transfer of the technology hastened its spread. Even if the technology had not been transferred to an overseas subsidiary, it might have leaked out about as quickly. According to executives of the firms in the sample, in most cases the technology transfer had little or no effect on how quickly the technologies leaked out. However, in about one-fourth of the cases, they felt that the technology transfer probably hastened foreigners' access to these technologies by at least 2½ years. Based on their estimates, there is no indication that the average effect has been greater in one industry than in another, but there is an indication that the transfer of a process technology tends to hasten its spread to a greater degree than does the transfer of a product technology.[21]

---

portional to Δ and independent of *t*. In other words, the Poisson process might be used as a simple model. If this model is applicable, the length of time before a technology leaks out has an exponential probability distribution. In fact, the observed distribution is not very different from such an exponential distribution, but the sample is too small to carry out a goodness-of-fit test. One assumption of this model is that the probability of a technology's leaking out to non-U.S. competitors is zero until it is transferred. This need not be the case, although none of the technologies in our sample leaked out before they were transferred. It is easy to relax this assumption. We are not saying that a model of this sort is the appropriate one; all we are proposing is that it is an interesting hypothesis to test.

21. On the average, according to the executives, these technology transfers hastened non-U.S. competitors' access to the technology by 3.1 years for processes and by 0.4 years for products. Using the Dixon-Massey test, which does not assume that the standard deviations are equal (but does assume normality), this difference is statistically significant. There is no contradiction between the assumption of normality here and the simple model in

To test some of these propositions, we carried out a χ^2 test to determine whether the lag between the time when a technology was first used in the United States and the time when non-U.S. competitors had access to it was affected by how soon after its first use in the United States it was transferred to an overseas subsidiary. The results indicate that the latter variable has no statistically significant effect on the lag. Thus, our data seem to bear out the opinions of the firms that, while their transfer to an overseas subsidiary may have hastened substantially the spread of a relatively small number of these technologies, it had little or no impact on the spread of most of them.[22]

The most frequent channel by which the technology "leaked out" was reverse engineering, according to the firms. That is, non-U.S. competitors took apart and analyzed the new or modified product to gain insights into the relevant technology. Also, information provided by patents was judged to play a significant role; and non-U.S. firms sometimes gained access to the technology by hiring away personnel employed by the subsidiary. As pointed out above, non-U.S. firms frequently could have obtained the technology in these or other ways if the transfer had not occurred. (For example, in some cases, they might have hired away personnel employed by U.S. firms, or they might have obtained the technology from the United States or another source in other ways.) However, in those cases where the technology transfer hastened their access to the technology, they often would not have been stimulated to do so, either because they would not have been aware of the technology's existence, or because they would not have felt it was profitable to try to obtain the technology in these ways.[23]

Richard Caves has pointed out: "Most of our knowledge about transfers of technology through the multinational firm is anecdotal and incomplete. What has been the experience of foreign subsidiaries of U.S. firms with the successful imitation of their technologies by foreign competitors?"[24] To help answer this question, each firm in our sample esti-

---

footnote 20 (or between the assumption and the data). Whereas we were concerned in footnote 20 with how rapidly the technology leaked out, here we are concerned with how much quicker it leaked out because of its being transferred to an overseas subsidiary.

22. The technologies in the sample were divided into two groups, depending on whether or not they were transferred to an overseas subsidiary within five years of their first use in the United States. They were also divided into two groups, depending on whether or not they became known to some non-U.S. competitor within eight years after their first use in the United States. The value of χ^2 for the resulting 2×2 contingency table (after Yates's correction was applied) was 1.65, which is far from statistically significant. This result suggests that the relationship between these two factors is not very strong.

23. It is worth adding that the non-U.S. competitors that have gained access to the technology often are not headquartered in the country where the overseas subsidiary is located. In this sample, they were headquartered most often in West Germany.

24. Caves, *op. cit.*, pp. 37–38.

TABLE 2.7

*Estimated Effect of Technology Transfer on
How Soon the First Non-U.S. Competitor Introduced
a Competing Product or Process,
Twenty-five Transfers of Technology*[a]

ACCELERATION IN INTRODUCTION OF COMPETING PRODUCT OR PROCESS (YEARS)	NUMBER OF TECHNOLOGY TRANSFERS
Zero and under 0.5	12
0.5 and under 2.5	4
2.5 and under 5.5	7
5.5 and under 10.5	1
10.5 and over	1

[a] Data on this score could not be obtained for one technology in the sample; thus, the total number included here is twenty-five.

mated the effect of the technology transfer on how rapidly each of its products or processes based on the technologies in the sample was imitated. As would be expected on the basis of this section's previous results, it appears that in most cases there was little or no effect. However, in about one-third of the cases, the product or process was imitated at least two and a half years earlier than the firms would have expected if the technology had not been transferred to the overseas subsidiary (Table 2.7). Once again, there is an indication that this effect tended to be greater for processes than for products.

7. Benefits to Non-U.S. Users and Suppliers

In a previous section, we described some of the concerns that have been expressed by host countries regarding international technology transfer via multinational firms. Despite these concerns, it is obvious that the transfer of technology by U.S.-based firms to their overseas subsidiaries may result in considerable benefits to the host (and other) countries. Yet despite the importance of measuring the size of these benefits, practically no quantitative information is available on this score.[25] In this section, we present some rough estimates of this sort.

The transferred technology can have benefits of various sorts. One

25. However, there have been a number of interesting and important studies, notably Brash, *op. cit.* (cited in Chapter 1, footnote 41); Richard Caves, "Multinational Firms, Competition, and Productivity in Host-Country Industries," *Economica* 41 (1974):176–93; and A. Safarian, *op. cit.* (cited in Chapter 1, footnote 41). Note that, when we refer in this section to non-U.S. users and suppliers, foreign subsidiaries of U.S.-based firms are excluded.

type of benefit is a reduction of the costs of non-U.S. *users* of the products or processes based on the transferred technology. For each technology in our sample (described in section 6), data provided by the U.S.-based firm were employed to estimate how much non-U.S. users of the goods produced with, or based on, this technology saved as a result of the technology transfer. These savings measure the cost reductions to non-U.S. users of these goods over and above what would have occurred if the technology had not been transferred (but the relevant technology and goods were perhaps available from the United States or elsewhere). They include savings due to on-the-spot training of users in the utilization of products and processes based on the technology, on-the-spot provision of complementary technology to users, and lower costs of producing and distributing the goods based on the technology, among other things. Estimates were made for each year subsequent to the transfer of the technology. Since only those savings that could be quantified with reasonable accuracy are included, the estimates are almost certainly conservative. For almost 80 percent of the technologies, savings of this sort occurred; the mean discounted saving (for the twenty-four technologies for which data are available) was about $13 million.[26]

Another way that the transferred technology can benefit non-U.S. economies is by cutting the costs of non-U.S. firms that *supply* inputs used to produce the product or process based on the transferred technology. For each technology in our sample, data provided by the U.S.-based firm were employed to estimate how much non-U.S. suppliers saved each year as a result of the technology transfer. Such savings often resulted from training programs and the transfer to suppliers of know-how that reduced the cost of producing components or other inputs used by the subsidiary in applying the transferred technology. As would be expected, savings to suppliers occurred less frequently than savings to users. For about 40 percent of the technologies, savings of this sort accrued to suppliers, the mean discounted saving (for all twenty-four technologies for which data are available) being about $4 million.[27] These

26. Throughout this section, when we estimate the savings due to each technology transfer, the comparison is with the costs of the users or suppliers if the transfer had not taken place. In many cases, if the transfer had not occurred, the technology could have been licensed from the U.S.-based firm or provided by other sources, or a different technology could have been used. The discount rate used here (and throughout this section) is 10 percent. Exchange rates were used to convert the savings into U.S. dollars. In five of the cases, the discounted savings to users were zero; in seven cases, they were positive but less than $1 million; in eleven cases, they were between $1 million and $100 million; and in one case, they were over $100 million.

27. In fourteen of the twenty-four cases, the discounted savings to suppliers were zero; in three cases they were positive but less than $1 million; and in seven cases, they were between $1 million and $100 million.

Whether or not the technology transfer resulted in cost reductions to non-U.S. users

estimates, like those in the previous paragraph, are almost certainly conservative, since only those savings that could be documented and quantified with reasonable accuracy are included.

How large have been the total resource savings per year to non-U.S.
users and suppliers due to technology transfer of this sort by the firms
in our sample? To obtain a rough lower bound on this amount, we calculated the annual cost reduction (to users and suppliers combined)
resulting from each of the jth firm's technology transfers in the sample.
For the ith technology transfer by the jth firm, this annual saving is
denoted by c_{ij}, and the number of years this saving continues is denoted
by t_{ij}. Then we obtained an estimate from each firm in our sample of the
annual number of technology transfers (during 1968–77) that resulted
in total savings to non-U.S. users and suppliers that were at least as large
as those resulting from the average technology transfer of this firm
included in our sample; the result for the jth firm is denoted by N_j. In a
steady-state condition where N_j technology transfers of this sort occur
each year in the jth firm, a lower bound on the total savings per year to
non-U.S. users and suppliers due to this firm's technology transfer is

$$(2.2) \qquad g_j = \frac{N_j}{m_j} \sum_{i=1}^{m_j} c_{ij} t_{ij},$$

where m_j is the number of technology transfers by the jth firm included
in the sample.[28] Inserting estimates of c_{ij}, N_j, and t_{ij} obtained from these
firms into this equation, we find that the sum of g_j for the firms in our
sample is about $633 million per year.

Based on this sample, a firm's value of g_j seems, on the average, to

---

or suppliers, it may have provided them with significant new information or new production capabilities. According to the U.S.-based firms, significant information of this sort was
transmitted to non-U.S. users in about two-thirds of the cases and to non-U.S. suppliers in
about one-third of the cases.

28. The total saving in a particular year (say year T) is equal to the sum of the savings
due to each of the technology transfers that still have an effect in year T. Consider a steady-
state situation where m_j transfers (with the values of c_{ij} and t_{ij} in the sample) occur each
year. For simplicity, assume that t_{ij} always assumes integer values of 1 or more. The total
saving in year T due to transfers in year T equals $\sum_{i=1}^{m_j} c_{ij}$, since t_{ij} in all cases is at least 1; the
total saving in year T due to transfers in year T-1 equals the sum of the c_{ij} for transfers
where t_{ij} is 2 or more; the total saving in year T due to transfers in year T-2 equals the sum
of the c_{ij} for transfers where t_{ij} is 3 or more; and so on. Thus, the total saving in year T
due to *all* transfers equals $\sum_{i=1}^{m_j} c_{ij} t_{ij}$. If N_j rather than m_j transfers occur per year (but if the
values of c_{ij} and t_{ij} remain the same), the total saving in year T would be $\frac{N_j}{m_j} \sum_{i=1}^{m_j} c_{ij} t_{ij}$. Based
on the past history and performance of these firms, the assumption of a steady-state situ-
ation of this sort does not seem unrealistic.

be (approximately) proportional to the total annual sales of its overseas subsidiaries, which, of course, is not surprising. If this sample of firms is representative, a lower bound for the total annual savings to non-U.S. users and suppliers due to all technology transfer of this sort by U.S.-based firms can be obtained by dividing $633 million by the proportion of all annual sales of U.S. firms' overseas subsidiaries accounted for by the firms in our sample. The result is approximately $35 billion, or about 1 percent of the total annual output of countries outside the United States.[29] Because it is based on a relatively small sample, this result should be treated with a great deal of caution. Moreover, it should be stressed once more that this is a lower bound.[30]

8. Effects on Technological Capabilities and Innovative Performance of British Firms

When U.S.-based firms transfer technology to overseas subsidiaries, there is likely to be an impact on many firms in the host countries (and elsewhere). But very little is known about the kinds of non-U.S. firms that are most affected. To obtain some information on this score, we focused attention on the United Kingdom, the country (outside North America) where direct investment by U.S.-based firms has been greatest. Seventy of the largest firms in the United Kingdom provided an assessment of the extent to which their technological capabilities during 1968–77 were affected by technological transfers by U.S.-based firms to their overseas subsidiaries. This information was obtained largely through correspondence and interviews.[31] Although we recognize that such assessments may be biased, we feel that the results are of interest.

29. When g_j is regressed on the jth firm's 1976 sales of its overseas subsidiaries, the slope, but not the constant, of the regression equation is statistically significant.

The 1970 aggregate gross domestic product (in noncentrally planned economies) outside the United States was about $2.4 trillion, according to I. Kravis, A. Heston, and R. Summers, "Real GDP per Capita for More Than One Hundred Countries," *Economic Journal* 88 (1978):215–42. Obviously, this figure is crude, but sufficient for present purposes. The figure of 1 percent is obtained by dividing $35 billion by $2.4 trillion.

30. Also, these estimates pertain only to the effects on users and suppliers, which are not the only affected parties. Of course, the statement in the text depends on the sample of firms being representative. It is always possible that sampling errors could offset the conservative bias we injected into the estimates.

31. The sample of seventy firms was selected from *Jane's Major Companies of Europe.* It included a wide variety of manufacturing industries: chemicals, electrical equipment (including electronics and computers), transportation equipment, general engineering, paper, textiles, food, commodities, building materials, construction, metals, oil and gas, beer, wine, and spirits, and alcohol. Each firm filled out a questionnaire concerning this topic. Then interviews were carried out with the managing director (or other major offi-

TABLE 2.8

Percent of Firm's New Products and Processes Introduced
(or Introduced More Quickly) Because of the Transfer of Technology by
a U.S.-based Firm to an Overseas Subsidiary,
Sixty-eight British Firms, 1968–77

PERCENT OF NEW PRODUCTS	NUMBER OF FIRMS	PERCENT OF NEW PROCESSES	NUMBER OF FIRMS
Zero	30	Zero	32
0.1 to 4	28	0.1 to 4	29
5 to 9	8	5 to 9	5
10 to 25	2	10 to 25	2
More than 25	0	More than 25	0
Totals[a]	68		68

[a] Data on this score could not be obtained from two firms in the sample; thus, the total number included here is sixty-eight.

About two-thirds of these firms felt that their technological capabilities during this period were raised (to some extent at least) by technology transfers of this sort. But only 20 percent felt that this effect was of importance.[32] The proportion regarding it as important was higher in the electrical-equipment (including electronics and computers) and general-engineering industries than in the others. Technology transfers of this sort seem to have had some effect on the innovative activities of most of these British firms. As shown in Table 2.8, over half of these firms felt that at least some of their products and processes had been introduced, or introduced more quickly, because of the transfer of a new product or process by a U.S.-based firm to an overseas subsidiary. But only 10 to 15 percent of these firms felt that 5 percent or more of their new products or processes were affected in this way. The percent of a firm's new products or processes influenced in this way tended to be larger in the electrical equipment (including electronics and computers) and oil and gas industries than in the others.

These data, despite their obvious imperfections, enable us to test for the first time two hypotheses that are often encountered in this area. The first hypothesis, put forth by Caves and others, is that non-U.S.

cial) of a subsample of these firms. Also, interviews were carried out with officials of overseas subsidiaries of U.S. firms.

32. Initially the firms were permitted to use whatever definition of "important" that they regarded as appropriate. Then they were asked to specify what they meant by this term. Based on the replies, it appears that they generally regarded any benefit that enhanced profits by one percent or more as important. Thus, their definition seemed rather liberal. There is no evidence of substantial differences among firms in the definition used.

firms in industries where subsidiaries of U.S. firms account for a large share of the local market are affected more by this kind of technology transfer than non-U.S. firms in industries where such subsidiaries account for a small share of the local market.[33] This hypothesis is by no means trivial since the amount and importance of technology transferred from the United States to the local market may bear little or no relationship to the subsidiaries' share of the local market. The second hypothesis stems from work at OECD and elsewhere, which indicates that the U.S. technological lead has tended to be greater in relatively R-and-D–intensive industries than in others. This hypothesis states that, if the share of the local market held by the subsidiaries of the U.S. firms is held constant, the impact tends to be greater on non-U.S. firms in industries where a relatively large percentage of sales is spent on research and development than in industries where a relatively small percentage of sales is spent for this purpose.

Each of the dependent variables described in Table 2.9 is a measure of the extent of the perceived impact on the ith British firm of technology transfer of this sort. To test these hypotheses, we regressed each of these dependent variables on the proportion of assets in the ith industry held by subsidiaries of the U.S. firms (M_i) and on the percent of sales spent on research and development by the ith industry (R_i). In addition, since a firm's size and profitability may be associated with its perception of the extent to which it is affected by technology transfer of this sort, we included the ith firm's sales (S_i) and its profit as a percent of sales (π_i) as additional variables in each regression. The signs of all the regression coefficients, shown in Table 2.9, are in accord with both hypotheses, but whereas the coefficients of M_i are almost always statistically significant, this is not true of the coefficients of R_i. (Also, the perceived impact of technology transfers of this sort seems to be greater for larger firms than for smaller ones, and greater for less profitable firms than for more profitable ones, but these differences are seldom statistically significant.) Thus, these results provide much stronger support for the first than for the second hypothesis.[34]

33. Caves (cited in note 25) put forth this hypothesis in connection with Australia and Canada, but it seems applicable to the United Kingdom as well.

34. To prevent confusion, note that we are investigating the characteristics of the firms that seem to be most affected by technology transfer of this sort. We are not asserting that these characteristics (measured by the independent variables in Table 2.9) necessarily cause the impact to be greatest among firms possessing them. Also, note that, although there is some positive correlation between M_i and R_i, it is not high enough to bar us from estimating the regression coefficients of both variables. The data regarding S_i and π_i come from *Jane's Major Companies of Europe*. The data regarding M_i pertain to 1965 and come largely from Dunning, and those regarding R_i pertain to 1969 and come from Freeman. See Dunning, *Studies in International Investment, op. cit.* (cited in Chapter 1, footnote 40) and Christopher Freeman, *The Economics of Industrial Innovation* (London: Penguin, 1974).

TABLE 2.9

Regression Coefficients of M_i, R_i, π_i, and S_i

DEPENDENT VARIABLE[a]	INDEPENDENT VARIABLE[b]			
	M_i	R_i	π_i	S_i
p_i	1.051 (1.06)	0.0282 (1.13)	−0.0486 (2.16)	0.000113 (1.72)
u_i	2.160 (2.44)	0.0773 (3.46)	−0.0177 (0.88)	0.0000296 (0.50)
v_i	17.62 (2.65)	0.121 (0.71)	−0.0966 (0.64)	0.0000180 (0.04)
w_i	22.59 (3.67)	0.0893 (0.57)	−0.0392 (0.28)	0.0000813 (0.20)

[a] The definitions of the dependent variables are as follows: p_i equals 1 if the ith firm indicated it received benefits from technology transfer due to purchases from subsidiaries of U.S. firms and zero otherwise; u_i equals 1 if the ith firm indicated it received benefits from technology transfer due to its supply of goods and services to subsidiaries of U.S. firms and zero otherwise; v_i is the percentage of the ith firm's new products that were introduced (or introduced more quickly) because of the transfer of technology by U.S. firms to their overseas subsidiaries; and w_i is the percentage of the ith firm's new processes that were introduced (or introduced more quickly) because of the transfer of technology by U.S. firms to their overseas subsidiaries.

When p_i is the dependent variable, $\bar{R}^2$ is 0.30; when u_i is the dependent variable, it is 0.46; when v_i is the dependent variable, it is 0.46; and when w_i is the dependent variable, it is 0.49.

[b] The definitions of the independent variables are the following: M_i is the proportion of assets in the ith firm's industry held by subsidiaries of U.S. firms; R_i is the percent of sales spent on research and development by the ith firm's industry; π_i is the profit (as a percent of sales) of the ith firm; and S_i is the sales (in millions of pounds) of the ith firm. Besides these independent variables, dummy variables were included to allow the intercept to vary from one broad industry group to another. For details, see footnote 34. The t-statistics are shown in parentheses under the regression coefficients.

9. Summary

There are many types of technology transfer. Vertical technology transfer occurs when information is transferred between basic research and applied research, between applied research and development, and between development and production. Horizontal technology transfer occurs when technology used in one place, organization, or context is transferred and used in another place, organization, or context. Material transfer consists of the export of a new material or product by one country to another. Design transfer consists of the transfer of designs, blueprints, and the ability to manufacture the new material or product in the recipient country. Capacity transfer occurs when the capacity to adapt the new item to local conditions is transferred. International technology transfer includes all of these types.

Although design transfer is a very important type, very little is known about international technology transfer of this type. Building on the work of Hufbauer and Tilton, we found that, in the plastics and semiconductor industries, there seemed to be a highly significant tendency for countries that spend relatively large amounts on R and D (in these industries) to have relatively short imitation lags. In the pharmaceutical industry, there appears to be no such tendency, perhaps because of regulatory considerations. In all three industries, when other factors are held constant, imitation lags tend to decrease with time. The effects of industrial concentration on a country's imitation lag seem to be mixed.

Based on a carefully selected sample of sixty-five technologies, the mean age of the technologies transferred by U.S.-based firms to overseas subsidiaries in developed countries during 1960–78 was about six years. For those transferred in this way to developing countries, the mean age was about ten years. For those transferred through licenses, joint ventures, and other channels other than subsidiaries, the mean age was about thirteen years. Between 1960–68 and 1969–78, there seemed to be a significant increase in the proportion of relatively new technology transferred to subsidiaries in developed countries. But for technologies transferred to subsidiaries in developing countries or for those transferred through channels other than subsidiaries, there appeared to be no such tendency, at least in this sample. These results are of interest since some observers worry that U.S. firms may have come to license to, or jointly exploit with, non-U.S. firms more and more of their newest technologies.

Among the technologies transferred to foreign subsidiaries, the mean lag between the technology transfer and the time when some non-U.S. competitor had access to the technology was estimated to be about four years. In most cases, the technology transfer seemed to have little or no effect on how quickly the technology leaked out, but in about one-fourth of the cases, it probably hastened foreigners' access to the technologies by at least two and a half years. The most frequent channel by which the technology leaked out was reverse engineering, according to the firms. In most cases, the technology transfer seemed to have little or no effect on how rapidly the relevant product or process was imitated, but in about one-third of the cases, it was imitated at least two and a half years earlier than would otherwise have been expected.

The technologies transferred by U.S.-based firms to their overseas subsidiaries can have benefits of various sorts to the host (and other foreign) countries. One type of benefit is a reduction of the costs of non-U.S. users of the products or processes based on the transferred technology. Another type of benefit is a reduction of the costs of non-U.S. firms that supply inputs used to produce the product or process based on the transferred technology. If our sample is representative, the annual saving to non-U.S. users and suppliers due to all technology transfer of this sort by U.S.-based firms seems likely to be at least $35 billion, or at least 1 percent of the total annual output of countries outside the United

Based on data obtained from about seventy of the largest firms in the United Kingdom, technological transfers by U.S.-based firms to their overseas subsidiaries have raised many British firms' technological capabilities to some extent, but an important effect has occurred in only a relatively few industries. In accord with Caves's hypothesis, industries where subsidiaries of the U.S. firms account for a large share of the local markets are affected more by this kind of technology transfer than other industries. There is also some apparent tendency for the impact to be greater in more R and D–intensive industries, but this tendency often is not statistically significant.

3 INTERNATIONAL TECHNOLOGY

TRANSFER: EFFECTS OF FOREIGN TRADE

ON U.S. R AND D

1. Introduction

Although economists have been aware for over a century that technological change and international technology transfer affect the pattern of world trade, there has been increasing emphasis on this factor in the past twenty years. Among the leaders in emphasizing this factor were the late Harry Johnson and Raymond Vernon. Both from case studies and statistical analyses, economists have found that U.S. industries that spend relatively large amounts on R and D are the ones that lead in manufactured exports, foreign direct investment, and licensing. Technological innovation and international technology transfer seem to have had a major influence on American exports, receipts from licenses, and direct investment abroad.

Despite the considerable recent interest in the effect of U.S. research and development on our foreign trade, little, if any, effort has been made to study the reverse flow of causal forces—namely, the effects of American foreign trade on our research and development activities. In 1973, the National Science Foundation convened a conference of leading economists that considered the question: What effects do international trade, foreign direct investment, and foreign licensing have on U.S. technological innovation? It was concluded that "in spite of the importance of this question there seems to be a complete void in our knowledge."[1] In this chapter, our purpose is to present some new findings bearing on this topic, as well as on the ways in which American firms

1. National Science Foundation, *The Effects of International Technology Transfers on U.S. Economy* (Washington: Government Printing Office, 1974), p. 5.

transfer their technology abroad. Although our findings are subject to a variety of limitations, they seem to represent some of the first systematic evidence related to these topics.

2. Returns from New Technology Utilized Abroad

To obtain the required data, a sample of thirty firms was chosen. This sample was composed of two parts, the first containing twenty firms in the fabricated metal products, machinery, instruments, chemical, textile, paper, and tire industries, the second containing ten major chemical firms.[2] The firms in both subsamples tend to be rather large. (Half of the firms in the ten-firm subsample had 1974 sales exceeding $1 billion, and over one-third of the firms in the twenty-firm subsample had 1974 sales exceeding $500 million.) The members of both subsamples are quite representative of all firms in their industries with regard to the percentage of sales devoted to research and development. Each of the firms in the sample was asked to provide detailed data concerning the extent to which foreign markets or foreign utilization contributed to the expected returns from each of its 1974 R and D projects.[3] Also, data were gathered concerning the characteristics of each firm's R and D portfolio, and the percentage of its sales derived from abroad. In general, the data were obtained from senior R and D executives and from officials involved

2. The basic data for the first subsample were the primary responsibility of Anthony Romeo. For this subsample, we collected data for all R and D projects, inside or outside the United States. The firms in this subsample were chosen more or less at random from major manufacturing firms in the southern New England and the Middle Atlantic states. The basic data for the second subsample were the primary responsibility of Samuel Wagner. For this subsample, we collected data only for R and D projects carried out in the United States. The firms in this subsample were chosen more or less at random from major chemical firms in the East. In this chapter, the "chemical" industry includes drugs and petroleum refining.

3. There are several ways that a firm can transfer and exploit its technology abroad. It can utilize the new technology in foreign subsidiaries, export goods that are based on the new technology, license the new technology to others who will use it abroad, or engage in joint ventures with others to use it abroad. The firms in the sample were asked to provide data concerning the percentage of expected profits from their R and D projects that were anticipated to come from any of these four sources. Obviously, such estimates could only be rough, but they seem sufficiently precise for present purposes. Specifically, for the ten-firm subsample, we asked each firm for the frequency distribution of R and D projects (weighted by size of expenditure) by this percentage. For the twenty-firm subsample, we asked each firm only for the *average* percentage of expected profits anticipated to come from these sources, since this reduced the burden imposed on the responding firms— and allowed us to go into more detail concerning other questions (for example, those taken up in section 4 and the latter part of section 3).

with the firm's international operations. These data were gathered over a period of months and were prepared with care by these executives.

If we lump all kinds of R and D projects from all of these firms together, how important, on the average, do foreign markets or foreign utilization bulk in the expected returns from these R and D projects? Although the two subsamples are entirely independent, they provide very similar answers to this question. In the chemical subsample, about 29 percent of an R and D project's returns, on the average, were expected to come from foreign sales or utilization. In the twenty-firm subsample,[4] about 34 percent of an R and D project's returns, on the average, were expected to come from these sources. Of course, averages of this sort must be viewed with caution, because they conceal a great deal of variation and are influenced by the industrial (and other) characteristics of the sample. But they provide a reasonable starting point for the analysis.[5]

Going a step further, let's disaggregate the results to the firm level and see how great the interfirm variation is and how it can be explained. In both subsamples, there is a large amount of interfirm variation: the average percentage of an R and D project's returns expected to come from foreign sales or utilization ranges from 0 in some firms to 50 or 60 in other firms. Two hypotheses may help to explain these interfirm differences. First, these differences are likely to reflect the fact that some firms, because of the nature of their product lines, their history, and their management, make a much larger percentage of their current sales overseas (through exports or sales of foreign subsidiaries) than do other firms. One would expect that such firms would tend to gear their R and D programs more heavily to foreign markets and utilization than other firms. Second, besides reflecting differences in the percentage of a firm's current sales that come from abroad (which may depend on *past* levels of R and D expenditures), these differences may also reflect differences in *present* levels of R and D spending. Since more R and D–intensive

4. One firm had to be omitted because it could not provide usable information on this score, so this figure really pertains to nineteen firms.

5. Of course, it is not surprising that the average for the twenty-firm subsample is somewhat higher than for the chemical subsample. As noted above, the former subsample includes R and D projects carried out overseas, whereas the latter does not. Since R and D projects carried out overseas typically are geared closely to foreign operations and markets, the inclusion of such projects would be expected to raise the average. But since such projects account for only about 9 percent of the total in this sample of firms, they cannot raise the average by more than about 6 percentage points. Why? Because even if the returns from every project carried out overseas come entirely from abroad, the average percentage of returns coming from abroad (including both projects carried out overseas and in the United States) would equal $0.91 \times P + 0.09 \times 100$, where P is the percentage of returns coming from abroad for projects carried out in the United States. And since we know that $0.91 \times P + 0.09 \times 100$ equals 34, it follows that P must be at least 28.

TABLE 3.1

Regression Equations to Explain Variation
(Among Firms or Product Lines)
in the Percentage of R and D Projects' Returns
Expected to Come from Abroad

REGRESSION COEFFICIENTS[a]

SUBSAMPLE[b]	CONSTANT	A	X	$\bar{R}^2$	n
20-firm	−4.04	0.66	288	.71	19
	(1.03)	(4.76)	(2.40)		
10-firm	6.07	0.44	6.4	.18	53
	(1.80)	(3.68)	(0.57)		

[a] The independent variables are defined as follows: A is the percentage of sales from abroad in 1974, and X is the ratio of R and D to sales in 1974. In the twenty-firm subsample, these variables pertain to the entire firm: in the ten-firm subsample, to an individual product line. The t-statistics are shown in parentheses, here and in subsequent tables.
[b] In the twenty-firm subsample, the dependent variable is the firm's average percentage of its 1974 R and D projects' returns expected to come from abroad; in the ten-firm subsample, it is the average percentage of the returns from R and D projects from a particular product line of a particular firm expected to come from abroad. These product lines are broadly defined, reflecting the fact that these firms produce a very wide variety of products.

industries seem to do more exporting, investing abroad, and licensing abroad than other industries, one might suspect that more R and D–intensive firms may expect a higher proportion of the returns from their R and D to come from abroad.

To test these hypotheses, we carried out a regression analysis of the data in each subsample. The results, shown in Table 3.1, provide significant evidence in support of the hypothesis that the relative importance of foreign markets and utilization in the returns from a firm's R and D program is directly related to the extent to which it depends on foreign sources for its current sales. Both subsamples also suggest that, holding constant a firm's percentage of sales from abroad, the relative importance of foreign markets and utilization in the returns from a firm's R and D program is greater in more research-intensive firms. (However, the latter tendency is statistically significant only in the twenty-firm sample.) Moreover, in the twenty-firm subsample, these hypotheses can explain about 70 percent of the interfirm variation in the relative importance of foreign markets and utilization in the expected returns from a firm's R and D program.[6]

Having looked at interfirm differences, let's turn to differences among types of R and D projects in the relative importance of foreign markets

6. The low value of $\bar{R}^2$ in the ten-firm subsample is due in part to the fact that the data on R and D contain arbitrary allocations of some R and D projects that affect more than one product line.

TABLE 3.2

Distribution of R and D Projects
(Weighted by Size of Expenditure),
by Estimated Percentage of Profit from Foreign Sales
or Utilization, Ten Major Chemical Firms, 1974

PERCENTAGE OF EXPECTED PROFIT FROM FOREIGN SALES OR USE	RESEARCH PROJECTS	DEVELOPMENT PROJECTS
Less than 10	25	34
10–24	14	15
25–49	34	27
50–74	26	24
75–89	0	0
90 or more	0	1
Total[a]	100	100
Mean Percentage	32	29

Source: See section 2.
[a] Because of rounding errors, the individual figures may not sum to the total.

and utilization. The very detailed data required to shed light on this topic were gathered only from the ten-firm chemical sample. Table 3.2 shows that, among both research projects and development projects, there is a great deal of variation on this score. On the average, research projects seem to have a somewhat larger share of their returns come from abroad than development projects, which seems reasonable (since the results of research projects probably tend to be less specific to the American market than development projects done here).[7] R and D projects aimed at new products seem to be the ones where foreign returns are expected to be most important, their foreign share being, on the average, about 40 percent.[8] According to executives of the firms, one principal reason why the returns from products come in larger measure from abroad than the returns from processes is that these firms are more hesitant to send overseas their process technology than their product technology because they feel that the diffusion of process technology, once it goes abroad, is harder to control. In their view, it is much more

7. However, there is considerable variation among firms in this regard. In four of the ten firms, development projects are expected to have a larger share of their returns come from abroad than research projects.

8. For the ten-firm sample, the mean estimated percentage of profit from foreign utilization or sales was 43 percent for R and D projects aimed at entirely new products, 22 percent for those aimed at product improvements, 21 percent for those aimed at entirely new processes, and 16 percent for those aimed at process improvements.

difficult to determine whether foreign firms are illegally imitating a process than a product.

3. Effects of Decreased Opportunities for International Technology Transfer on U.S. R and D Expenditures

In recent years considerable controversy has raged over the effects of direct investment abroad (and other channels of international technology transfer) on America's technological position. As pointed out in previous chapters, some observers are concerned that such investment may result in a reduction in our technological lead, since U.S. technology may tend to be transferred from our foreign subsidiaries to our foreign competitors. However, a point that is often ignored is that, if American firms could not establish foreign subsidiaries (or transfer technology abroad in other ways), they would not carry out as much research and development, with the result that our technological position would be weakened. Some economists, like Caves and Stobaugh, have recognized this point, but have cited the unfortunate fact that nothing is really known about the amount by which U.S. R and D expenditures would decline if U.S firms could not transfer their technology to their foreign subsidiaries or use other channels of international technology transfer.

In this section we present what seem to be the first published data on this score. To obtain these data, we asked the thirty firms in our sample to estimate how much their R and D expenditures would have changed in 1974 under two sets of circumstances: (1) that they could not utilize any new technology abroad in foreign subsidiaries; (2) that they could not utilize any new technology abroad in foreign subsidiaries, or by licensing the technology abroad, or by exporting new products or processes based on the technology, or by any other means.[9] (For the twenty-firm subsample, data were obtained for 1970 as well.)[10] Although answers to hypothetical questions of this sort must be treated with caution, the results should be of interest. Moreover, as we shall see, a comparison of these results with some earlier econometric findings suggests that, if anything, these results may be on the conservative side.

9. The second set of circumstances was chosen because it would suggest what would occur if, as some say, firms base their R and D decisions solely on the basis of expected domestic returns. The first set of circumstances was chosen to see how much substitutability exists among various channels of international technology transfer in the eyes of the firms. Neither of these sets of circumstances was chosen because there has been any serious attempt to institute public policies that would lead to so severe a limitation on the transfer of technology. In this connection, see Richard Caves's paper (cited in Chapter 2, footnote 1); and Raymond Vernon, "The Location of Economic Activity," in Economic Analysis and the Multinational Enterprise, ed. John Dunning (London: Allen and Unwin, 1974).

10. The 1970 data were collected only for the second set of circumstances.

Based on the firms' estimates, their R and D expenditures would have fallen significantly under each of the above sets of circumstances. Specifically, for the twenty-firm subsample, the estimated reduction would have been about 15 percent in 1974 if they could not utilize any new technology in foreign subsidiaries, and about 26 percent in 1974 (17 percent in 1970) if they could not transfer any new technology abroad by any means. For the 10-firm chemical subsample, the estimated reduction in 1974 would have been about 12 percent if they could not utilize any new technology in foreign subsidiaries, and about 16 percent if they could not transfer any new technology abroad by any means. Thus, the results obtained from the two (quite independent) subsamples are reasonably close.[11] Further, one can compare these results with what would be expected from an econometric model constructed some years ago by Mansfield: based on the finding in the previous section that about 30 percent of these firms' expected returns from their R and D projects stem from some form of international technology transfer, it appears that, if anything, these results underestimate the reduction in R and D expenditures that would result under these circumstances.[12]

11. Of course, the figure for the twenty-firm subsample is not entirely comparable with that for the chemical subsample, since the latter includes only R and D expenditures in the United States, whereas the former includes overseas R and D expenditures by these firms as well. If all overseas R and D expenditures would have been eliminated under these circumstances, then the R and D expenditures in the United States of the twenty-firm sample would have been reduced in 1974 by about 5 percent if they could not utilize any new technology in foreign subsididaries, and by about 15 percent if they could not transfer any technology abroad by any means. This follows from the fact that about 9 percent of these firms' R and D expenditures were carried out abroad. If the chemical subsample carried out about 10 percent of their total R and D abroad, then their total 1974 R and D expenditures (including foreign) would decrease by about 21 percent if they could not utilize any new technology in foreign subsidiaries and by about 24 percent if they could not transfer any technology abroad by any means. These calculations are obviously extremely rough, since they assume that all foreign R and D expenditures of these firms would cease.

12. Suppose that a firm expects about 30 percent of each R and D project's returns to come from abroad. Then, if the firm could not exploit its R and D results abroad, the expected rate of return from each project would be reduced by about 30 percent. Based on the model in Mansfield, *Industrial Research and Technological Innovation, op. cit.* (cited in Chapter 1, footnote 1), this would result in a percentage change in R and D expenditures equal to

$$[.7^{\frac{\bar{p}}{\bar{p} - p^*}} - 1] \times 100,$$

where $\bar{p}$ is the average rate of return from the firm's R and D projects (before it could no longer exploit its R and D results abroad) and p^* is the minimum estimated rate of return the firm will accept for R and D projects. Based on the data in Mansfield, $\bar{p} \div (\bar{p} - p^*)$ can be expected to be at least 1. Thus, one would expect the percentage reduction in R and D expenditures under these circumstances to be at least 30. Of course, the crudeness of these

What sorts of firms would cut their R and D expenditures most heavily under these circumstances? We would expect that the percentage reduction in R and D expenditures under these circumstances would be greatest among firms where foreign sales are a relatively large proportion of total sales, because, as indicated in the previous section, a relatively large share of the returns from such firms' R and D projects are expected to come from abroad. Thus, the inability to utilize or transfer technology abroad is likely to cut more deeply into the expected profitability of such firms' R and D projects, and to make a larger proportion of such firms' R and D projects economically unattractive. Also, we would expect that the percentage reduction in R and D expenditures under these circumstances would be greater for firms that spend a relatively large amount (in absolute terms) on R and D. Essentially, this is because indivisibilities may prevent firms that spend little on R and D (in absolute terms) from making as large a percentage reduction as bigger R and D spenders.[13] These hypotheses were tested against the data from both subsamples, using regression analysis. The results, shown in Table 3.3, indicate that the regression coefficients have the expected signs, and are statistically significant in most cases.

What sorts of R and D projects would be cut most heavily under these circumstances? To obtain some information on this score, we asked each member of our twenty-firm subsample to estimate (under the second set of circumstances described above) the percentage reduction that would have occurred in the amounts spent on basic research, applied research, and development, as well as on R and D aimed at new products, product improvements, new processes, and process improvements. The results indicate relatively little differences among the size of the percentage reductions for basic research, applied research, and development, although the reduction for basic research seems to be slightly larger than for the others. The big difference is between the percentage reduction for product R and D and process R and D: The percentage cut for the former would be about 35 percent, whereas for the latter it

---

estimates should be emphasized since the model is rough and no account is taken of the substantial variation among projects in the percentage of returns expected to come from abroad or of any possible correlation between this percentage and the project's value of p.

13. If a firm spends a great deal on R and D, certain parts of its R and D establishment may specialize in areas that are closely linked with overseas uses. Under the indicated sets of circumstances, they might no longer be profitable to maintain. However, if a firm spends relatively little on R and D, it may be difficult to close down any substantial part of its organization or facilities because to do so would put its R and D establishment below the minimum efficient size. For some discussion of the minimum economic scale of R and D establishments, see Chapter 5.

TABLE 3.3

*Regression Equations to Explain Variation (Among Firms or Product Lines)
in Estimated Percentage Reduction in R and D Expenditures
If New Technology Could Not Be Utilized Abroad in Foreign Subsidiaries* [a]

SUBSAMPLE[c]	CONSTANT	REGRESSION COEFFICIENTS[b]				$\bar{R}^2$	n
		u	E	A	r		
20-firm[d]	0.231	0.46	0.26	—	0.06	.69	20
	(0.06)	(4.17)	(0.87)		(1.46)		
10-firm	−7.71	—	—	0.24	0.49	.41	53
	(2.63)			(2.31)	(4.39)		

[a] The results for both subsamples are essentially the same as those shown here, if the dependent variable is the estimated percentage reduction in R and D expenditures if new technology could not be used abroad by any means.
[b] For the twenty-firm subsample, the independent variables are defined as follows: u is the percentage of a firm's sales from foreign subsidiaries in 1970 or 1974, E is the percentage of a firm's sales from exports in 1970 or 1974, and r is the firm's R and D expenditures (in millions of dollars) in 1970 or 1974. For the ten-firm subsample, the definitions are the following: A is the percentage of the sales of a particular product line of a particular firm that comes from abroad in 1974, and r is the firm's R and D expenditures (in millions of dollars in 1974). In the ten-firm sample, data were obtained concerning only A, not E and u separately.
[c] For the twenty-firm subsample, the dependent variable is the firm's estimated percentage reduction in R and D expenditures in 1970 or 1974; for the ten-firm subsample, it is the firm's estimated percentage reduction in R and D expenditures related to a particular product line in 1974.
[d] Besides the constant, three industry dummy variables are included. The first, which equals 1 if the firm is in the fabricated metal products industry and 0 otherwise, has a regression coefficient of −2.70; the second, which equals 1 if the firm is in the instruments industry and 0 otherwise, has a regression coefficient of −0.89, and the third, which equals 1 if the firm is in the chemical industry and 0 otherwise, has a regression coefficient of −7.65.

would be less than 5 percent. This, of course, is quite consistent with our finding in the previous section that a much larger share of the returns from product R and D (than from process R and D) is expected to come from abroad.

4. The Relationship between a Firm's Foreign Sales and the Size and Characteristics of Its R and D Portfolio

Are a firm's R and D intensity and the characteristics of its R and D portfolio related to the extent of its foreign sales? In other words, do firms with high ratios of foreign to total sales differ significantly from those with low ratios of this sort with regard to the percentage of sales devoted to R and D and the characteristics of their R and D portfolio? Gruber, Mehta, and Vernon have shown that industries that devote a relatively high percentage of sales to R and D tend to have high ratios of foreign to total sales. But holding industry constant, Thomas Horst found that only size of firm, not R and D intensity, was related to the extent of a firm's multinational involvement.[14] Our own results are in accord with this finding in the sense that in neither subsample was there any significant correlation between the percentage of a firm's sales that stem from abroad and its ratio of R and D expenditures to sales, when industry is held constant.[15]

Although previous investigators have looked at the relationship between R and D intensity and the extent of a firm's foreign sales, no attention has been devoted to the relationship between the kinds of R and D carried out by a firm and the extent of its foreign sales. This is unfortunate, since R and D is by no means homogeneous, and the nature of a firm's R and D activities may for many purposes be as important as their total amount. Two measures that frequently are used to characterize the extent to which a firm's R and D expenditures are for relatively fundamental or long-term work are (1) the percentage of the firm's R and D expenditures that goes for basic research and (2) the percentage of its R and D projects that extends for two years or more. One might argue that firms that rely relatively heavily on foreign sales may find it profitable to do somewhat more fundamental, long-term R and D than other firms, because such R and D is more likely to find application in some part of their more heterogeneous market and operations. In a sense, this argument is like that presented by Richard Nelson regarding diversified firms.[16]

To test this hypothesis, we calculated the regressions shown in Table 3.4. If time, industry, and size of firm (as measured by sales) are held constant, the results suggest that firms that rely more heavily on foreign sales tend to devote more of their R and D expenditures to basic research and to longer-term projects. Further, the results suggest that they tend to regard their R and D expenditures as being more profitable than do

14. William Gruber, Dileep Mehta, and Raymond Vernon, "The R and D Factor in International Trade and International Investment of U.S. Industries," *Journal of Political Economy* 75 (1967):20–37; and Thomas Horst, "Firm and Industry Determinants of the Decision to Invest Abroad: An Empirical Study," *Review of Economics and Statistics* 54 (1972):258–66.

15. For the twenty-firm subsample, we regressed each firm's percentage of sales devoted to R and D (in 1974 and 1970) on its sales, the percentage of its sales from exports, the percentage of its sales from foreign subsidiaries, and time and industry dummies. No variable was significant. For the ten-firm chemical subsample, there is no significant correlation between a firm's percentage of sales devoted to R and D and the percentage of its sales from abroad. Also, for thirty-five firms chosen from *Fortune*'s 500, we obtained data concerning each firm's percentage of sales devoted to R and D in 1974 and the percentage of its sales from abroad in that year. Then we regressed the former variable on the latter variable and on industry dummies. Only the industry dummies were statistically significant. The thirty-five firms are described in Chapter 5.

16. Richard Nelson, "The Simple Economics of Basic Scientific Research," *Journal of Political Economy* 67 (1959):297.

TABLE 3.4

*Regression Equations Showing the Relationship between
a Firm's Percentage of Sales Coming from Abroad and
Selected Characteristics of Its R and D Expenditures*

DEPENDENT VARIABLE[b]	CONSTANT[c]	REGRESSION COEFFICIENTS[a]				$\bar{R}^2$	n[d]
		S	E	u	T		
B	−1.29 (1.90)	0.0012 (2.92)	0.323 (4.92)	0.027 (1.26)	−1.28 (1.94)	.53	40
L	4.25 (0.62)	0.0094 (2.23)	3.87 (5.29)	0.13 (0.60)	−7.78 (1.12)	.65	36
π	0.45 (4.59)	−0.000073 (1.26)	0.015 (1.67)	0.009 (2.03)	−0.03 (0.34)	.48	25

[a] The definitions of the independent variables are as follows: S is the firm's sales (in millions of dollars) in 1970 or 1974, T is a dummy variable that is 1 in 1974 and 0 in 1970, and u and E are defined in Table 3.3.

[b] The definitions of the dependent variables are as follows: B is the percentage of the firm's R and D expenditures that went for basic research (as defined by the National Science Foundation) in 1970 or 1974, L is the percentage of the firm's R and D projects in 1970 or 1974 that extended for two or more years, and π is the median rate of return estimated by the firm for its R and D projects in 1970 or 1974 (if the projects are successful). The estimates of π are only rough.

[c] Besides the constant, there is an industry dummy (which equals 1 if the firm is in the fabricated metal products industry and 0 otherwise) in the equation for B; its regression coefficient equals 1.72. In the equation for L, there is an industry dummy (which equals 1 if the firm is in the instruments industry and 0 otherwise); its regression coefficient equals 23.5. In the equation for π, there are three industry dummies: the first (which equals 1 if the firm is in the fabricated metal products industry and 0 otherwise) has a regression coefficient of −0.28, the second (which equals 1 if the firm is in the instruments industry and 0 otherwise) has a regression coefficient of −0.26, and the third (which equals 1 if the firm is in the chemical industry and 0 otherwise) has a regression coefficient of −0.32. Other industry dummies were tried but were not close to being significant.

[d] The number of observations is less than forty in the regressions explaining L and π because not all firms could provide data concerning L and π.

other firms. To some extent, these results may reflect the fact that firms with large foreign markets find R and D, and particularly basic and longer-term R and D, relatively profitable. But the line of causation may also run in the reverse direction, since firms that tend to focus their R and D activities on more basic, long-term projects and that regard R and D as being relatively profitable may be more likely to establish themselves as international technological leaders and thus to gain large foreign sales. All that can be done here is to suggest that an association of this sort may exist; the construction of a multiequation model to help sort out this identification problem lies outside the scope of this chapter.

5. *Channels of International Technology Transfer*

While earlier sections of this chapter have indicated that many industrial R and D projects are carried out with the expectation that a consid-

TABLE 3.5

Percentage Distribution of R and D Projects,
by Anticipated Channel of International
Technology Transfer, First Five Years after
Commercialization, Twenty-three Firms, 1974 [a]

CHANNEL OF TECHNOLOGY TRANSFER

CATEGORY	FOREIGN SUBSIDIARY	EXPORTS	LICENSING	JOINT VENTURE	TOTAL[b]
All R and D projects:					
16 industrial firms	85	9	5	0	100
7 major chemical firms	62	21	12	5	100
Projects aimed at[c]					
Entirely new product	72	4	24	0	100
Product improvement	69	9	23	0	100
Entirely new process	17	83	0	0	100
Process improvement	45	53	2	1	100
Projects where estimated rate of return (if commercialized) is[d]					
Less than 20%	36	19	38	7	100
20% to 39%	46	29	19	5	100
40% or more	100	0	0	0	100

Source: See section 2.

[a] Only projects where foreign returns are of some expected importance (more than 10 percent of the total for the sixteen industrial firms and 25 percent of the total for the seven chemical firms) are included.

[b] Because of rounding errors, percentages may not sum to 100.

[c] Only six chemical firms could be included.

[d] Only four chemical firms could be included.

erable portion of their returns will come from abroad, no attention has been focused as yet on the channels by which these firms intend to effect these international transfers of technology. As indicated in previous chapters, there is a variety of such channels (foreign subsidiaries, exports, licensing, and joint ventures), but, as many researchers have pointed out, very little is known about the extent to which firms of various sorts use each of these channels. Table 3.5 shows, for our sample of firms, the percentage of all R and D projects (for which foreign returns were estimated to be of substantial importance)[17] where the principal channel (in the first five years after the commercialization of the new technology) was anticipated to be of each type. The results, which are much the same in the two subsamples, indicate that foreign subsidiaries are expected to

17. For the chemical subsample, we specified that more than 25 pcercent of total returns should come from abroad if a project is to be included. For the twenty-firm subsample, we specified that more than 10 percent should come from abroad if a project is to be included.

be the most frequently used channel, exports and licensing coming next, followed by joint ventures.

The great preponderance of cases where foreign subsidiaries are regarded as the principal channel during the first five years after commercialization is noteworthy because, according to the traditional view, the first channel of international technology transfer often is exports. Only after the overseas market has been supplied for some time by exports would the new technology be transferred overseas via foreign subsidiaries, according to this view. To some extent, our results may reflect an increased tendency for new technology to be transferred directly to overseas subsidiaries or a tendency for it to be transferred more quickly to them (in part because more such subsidiaries already exist). Such tendencies have been observed in the pharamaceutical industry, where many new drugs developed by U.S. firms have been introduced first by their subsidiaries in the United Kingdom and elsewhere. Also, Baranson's study concludes that American firms in a variety of industries are more willing than in the past to send their most recently developed technology overseas. Since we lack data like those in Table 3.5 for earlier periods, we cannot really test whether these tendencies exist in our sample.

Table 3.5 also indicates that product innovations are more likely to be transferred abroad via foreign subsidiaries than process innovations, while process innovations are more likely to be transferred abroad via exports than product innovations. To some extent, this seems to be due to more reluctance to send new processes overseas than new products. (Recall section 2.) Thus, there seems to be a greater tendency among processes than products for the technology to be exploited abroad through exports, since this channel does not require the technology to leave home.[18] In addition, Table 3.5 indicates that firms are more likely to license innovations that are only marginally profitable than ones that are very profitable, and that they are more likely to transfer very profitable innovations via subsidiaries than by licensing. To some extent, this may reflect the innovator's reluctance to provide foreign producers with information and legal rights, since this may hasten the day when the innovator will face competition in its own domestic market or elsewhere.

---

18. One might also suppose that process innovations would be more likely to be transferred via licensing than product innovations. For example, as Caves has pointed out, process technology can probably be transmitted more easily at arm's length from one firm to another, and the licensor can more readily drive a reasonable bargain without detailed knowledge of the specific market conditions facing the foreign licensee. However, such a tendency does not show up in Table 3.5, perhaps because of the smallness of the sample. See Caves, *op. cit.;* and Robert Wilson, "The Sale of Technology through Licensing," Ph.D. diss., Yale University, 1975.

Further, it is more likely to be worth the innovator's while to invest the resources to produce the innovation abroad in its own subsidiaries if the innovation appears very profitable. Beyond this, it may be that some innovations are more profitable because they can be exploited in foreign subsidiaries rather than licensed.

Changes over time would be expected in the principal channel by which a new technology is transferred abroad. In particular, as the technology grows older, there may be a tendency for exports to become a less important channel, since, as noted above, the innovator may supply foreign markets to a greater extent by foreign subsidiaries. Also, licensing may become more important because, as the technology becomes more widely known, foreign countries can take advantage of competition among technologically capable firms to obtain licenses rather than accept wholly owned subsidiaries. To see whether such tendencies exist in our sample, we obtained data of the sort shown in Table 3.5 for the second, rather than the first, five years after the commercialization of the new technology. In accord with our hypotheses, the results suggest that licensing is more important and exports are less important in the second five years than in the first.[19]

6. Summary

In our sample of firms, about 30 percent of an R and D project's returns, on the average, were expected to come from foreign sales or utilization. The relative importance of foreign markets and utilization in the returns from a firm's R and D program is directly related to the extent to which it depends on foreign sources for its current sales and to its ratio of R and D expenditures to sales. On the average, research projects seem to have a somewhat larger share of their returns come from abroad than development projects. R and D projects aimed at new

19. In the second five years, the percentage distribution of R and D projects by anticipated channel of international technology transfer was 60 percent (foreign subsidiaries), 15 percent (exports), 18 percent (licensing), and 6 percent (joint ventures). This distribution pertains to six of the chemical firms, and should be compared with the second row in Table 3.5. Of course, this comparison is clouded by the fact that one of the firms in the second row in Table 3.5 could not be included in this distribution. But if this firm is excluded from the second row of Table 3.5, the results are much the same. Note too that the change in the percentage attributable to foreign subsidiaries is quite small and of dubious significance, although it does have the expected sign.

Returning to the data for the first five years, there is a statistically significant tendency for larger firms to rely more heavily on foreign subsidiaries than smaller firms, and for smaller firms to rely more heavily on exports than larger firms. This would be expected, since larger firms are more likely already to have foreign subsidiaries, and to be in a better position to obtain the capital to establish new ones.

products seem to be the ones where foreign returns are expected to be most important, their foreign share being, on the average, about 40 percent. These firms are more hesitant to send overseas their process technology than their product technology, because they feel that the diffusion of process technology, once it goes abroad, is harder to control.

As pointed out in previous chapters, some observers are concerned that direct investment abroad may result in a reduction in our technological lead, since U.S. technology may tend to be transferred from our foreign subsidiaries to our foreign competitors. However, a point that is often ignored is that, if American firms could not establish foreign subsidiaries (or transfer technology abroad in other ways), they would not carry out as much research and development, with the result that our technological position would be weakened. For the firms in our sample, the estimated reduction in R and D expenditures would have been about 15 percent if they could not utilize any new technology in foreign subsidiaries and about 20 percent if they could not transfer any new technology abroad by any means. The percentage reduction would have been relatively high among firms where foreign sales are a relatively large proportion of total sales and where R and D expenditures (in absolute terms) are large. And it would have been much greater for product R and D than for process R and D.

When industry is held constant, there is no relationship between the percentage of a firm's sales that stem from abroad and its ratio of R and D expenditures to sales. But there does seem to be a relationship between the percentage of a firm's sales that stem from abroad and the characteristics of its R and D. If time, industry, and size of firm (as measured by sales) are held constant, the results suggest that firms that rely more heavily on foreign sales tend to devote more of their R and D expenditures to basic research and to longer-term projects. To some extent, this may reflect the fact that firms with large foreign markets find basic and longer-term R and D particularly profitable. Also, the line of causation may run in the reverse direction as well.

For our sample of firms, the principal channel of international transfer (in the first five years after commercialization) was foreign subsidiaries for about 70 percent of the R and D projects. This is noteworthy because, according to the traditional view, the first channel of international technology transfer often is exports. Only after the overseas market has been supplied for some time by exports would the new technology be transferred overseas via foreign subsidiaries, according to this view. Our results seem to reflect an increased tendency for new technology to be transferred directly to overseas subsidiaries or a tendency for it to be transferred more quickly to them (in part because more such subsidiaries already exist).

4 THE RESOURCE COSTS

OF TRANSFERRING

INDUSTRIAL TECHNOLOGY

1. Introduction

International technology transfer results in both benefits and costs. In Chapters 2 and 3, we discussed some of the benefits to both foreigners and Americans from the transfer of technology by U.S. firms to overseas subsidiaries. In the present chapter, we turn to the costs of transferring industrial technology. The starting point of our discussion is Kenneth Arrow's suggestion that the cost of communication, or information transfer, is a fundamental factor influencing the worldwide diffusion of technology.[1] The purpose of this chapter is to examine the level and determinants of the costs involved in transferring technology. The value of the resources that have to be utilized to accomplish the successful transfer of a given manufacturing technology is used as a measure of the cost of transfer. The resource cost concept is designed to reflect the ease or difficulty of transferring technological know-how from manufacturing plants in one country to manufacturing plants in another.

2. Technology Transfer and the Production of Knowledge

As pointed out in Chapter 1, the literature on technological change recognizes that it takes substantial resources to make a new process or

1. Arrow asks: "If one nation or class has the knowledge which enables it to achieve high productivity, why is not the other acquiring that information? . . . The problem turns on the differential between costs of communication within and between classes" (or nations). Kenneth Arrow, "Classificatory Notes on the Production and Transmission of Technological Knowledge," *American Economic Review* 59 (1969):33.

product feasible. However, it is common to assume that the cost of transferring the innovation to other firms is very much less, so that the marginal cost of successive applications is trivial compared to the average cost of research, development, and innovation. This paradigm is sometimes extended to international as well as domestic technology transfer.[2] Buttressing this view is a common belief, cited in Chapter 2, that technology is nothing but a set of blueprints that is usable at nominal cost to all. Nevertheless, it has been pointed out that generally "only the broad outlines of technical knowledge are codified by nonpersonal means of intellectual communication, or communication by teaching outside the production process itself." The cost of transfer, which can be defined to include both transmission and absorption costs, may therefore be considerable when the technology is complex and the recipient firm does not have the capabilities to absorb the technology. The available evidence is unfortunately very sketchy. From the case studies of Mueller and Peck, Arrow inferred that transfer costs must be high. From the Hall and Johnson study of the transfer of aerospace technology from the United States to Japan, it is not clear that this is true. E. A. G. Robinson believes that economists tend to exaggerate the size of transfer costs, while Edwin Mansfield and Christopher Freeman take the opposite view. The lack of compelling evidence is apparent, and the appeals for further research seem to be well founded.[3]

3. The Sample

The domain of this study is the transfer of the capability to manufacture a product or process from firms in one country to firms in another. In the language of Chapter 2, the transfers can be considered as horizontal, and in the design phase. Data on twenty-six fairly recent international technology transfer projects were obtained. The proprietary nature of much of the data meant that sampling costs were high, which in turn severely limited the size of the sample that could be collected. All twenty-six transfers were conducted by firms that were multinational in

2. "Transmission of technology between countries is assumed costless. Thus, it is possible for the country which owns the technology to operate a plant in a foreign country without any transfer of factors." C. Rodriguez, "Trade in Technical Knowledge and the National Advantage," *Journal of Political Economy* 93 (1975):122.

3. See Kenneth Arrow, "Comment," in National Bureau of Economic Research, *The Rate and Direction of Inventive Activity* (Princeton: Princeton University, 1962); Hall and Johnson, *op. cit.* (cited in Chapter 2, footnote 3); Christopher Freeman, "Research and Development in Electronic Capital Goods," *National Institute Economic Review* (1965); and the comments by Edwin Mansfield and E. A. G. Robinson in *Science and Technology in Economic Growth*, ed. B. Williams (London: Macmillan, 1973).

the scope of their manufacturing activity, although they varied considerably in sales value and R and D expenditures (1.2 to 12.5 percent of sales value). All had headquarters in the United States. The transferees were on the average much smaller and less research-intensive. In twelve instances they were wholly owned subsidiaries of the transferor, in eight instances the transferor and transferee were joint ventures partners, in four instances transfers were to wholly independent private enterprises, and the remaining three were to government enterprises. Table 4.1 shows that seventeen of the projects fall into a broad category that will be labeled "chemicals and petroleum refining." The remaining nine projects fall into a category which will be labeled "machinery." Table 4.1 also indicates the wide geographical dispersion of the transferees.

4. Definition of Technology Transfer Costs

An economic definition of transfer cost is developed below. The emphasis is on the resources that must be utilized to transfer technological know-how. Of course, royalty costs or rents must be incurred merely to secure access to the technology, but these costs are not the focus of attention of this paper.[4] In order to appreciate the import of the definition that will be presented, a distinction must first be made between two basic forms in which technology can be transferred. The first form embraces physical items such as tooling, equipment, and blueprints. Technology can be embodied in these objects. The second form of technology is the information that must be acquired if the physical equipment or "hardware" is to be utilized effectively. This information relates to methods of organization and operation, quality control, and various other manufacturing procedures. The effective conveyance of such "peripheral" support constitutes the crux of the process of technology transfer, and it typically generates the associated information flows. It is toward discovery of the cost of transfer of this "unembodied" knowledge[5] that our attention is directed.

Technology transfer costs are therefore defined as the costs of transmitting and absorbing all of the relevant unembodied knowledge. The costs of performing the various activities that have to be conducted to ensure the transfer of the necessary technological know-how will repre-

4. Many observers equate the cost of technology with royalty fees. See R. Hal Mason, "The Multinational Firm and the Cost of Technology to Developing Countries," *California Management Review* 15 (1973):5–13; and R. Gillette, "Latin America: Is Imported Technology Too Expensive?" *Science*, July 6, 1973, pp. 4–44.

5. *Unembodied knowledge* is the term used here to denote knowledge not embodied in capital goods, blueprints, and technical specifications, etc.

TABLE 4.1

Twenty-six Technology Transfer Projects:
Three-Digit ISIC Category and Transferee Location

| LOCATION | CHEMICALS AND PETROLEUM REFINING | | | MACHINERY | | | TOTAL |
	351: INDUSTRIAL CHEMICALS	353: PETROLEUM REFINERIES	356: PLASTIC PRODUCTS	381: FABRICATED METAL PRODUCTS MACHINERY AND EQUIPMENT	382: MACHINERY EXCEPT ELECTRICAL	383: ELECTRICAL MACHINERY, APPLIANCES, AND SUPPLIES	
Canada	1	1	0	1	0	0	3
Northern and western Europe	3	1	0	0	4	1	9
Australia	0	0	1	0	0	0	1
Japan	3	0	0	0	1	0	4
Eastern Europe	2	0	0	0	0	0	2
Latin America	3	0	0	0	0	1	4
Asia (excluding Japan)	0	1	0	0	0	1	2
Africa	1	0	0	0	0	0	1
Totals	13	3	1	1	5	3	26

sent the cost of technology transfer.[6] Clearly, a great many skills from other industries (e.g., design engineering) will be needed for plant design, plant construction, and equipment installation. However, not all of these skills will have to be transferred to ensure the success of the project. As defined, the costs of transfer clearly do not include all of the costs of establishing a plant abroad and bringing it on stream.

The definition of transfer costs presented at the conceptual level can be translated into operational measures by considering the nature of a given project activity. At the operational level the subset of project costs identified as transfer costs fall into four groups. The first group is the cost of pre-engineering technological exchanges. During these exchanges the basic characteristics of the technology are revealed to the transferee, and the necessary theoretical insights are conveyed. The second group of costs included are the engineering costs associated with transferring the process design and the associated process engineering[7] in the case of process innovations, or the product design and production engineering[8] in the case of product innovations. If the technology has already been commercialized,[9] transmission may simply involve transferring existing drawings and specifications with the minimum of modification. However, the process of absorption may be more difficult, requiring the utilization of considerable consulting or advisory resources. "Engineering" costs not falling into the specified categories[10] are excluded from transfer costs. The excluded engineering costs are essentially the plant or detailed engineering costs, net of advisory or consulting costs. This residual is assumed to correspond with routine drafting costs. Routine drafting is generally performed by technicians under the supervision of engineers. Drafting skills do not have to be transferred for the viability of the project to be assured. Accordingly, drafting is not considered to represent a transfer activity.[11]

6. All of the relevant costs are included, regardless of which entity initially or eventually incurs them.

7. See *McGraw-Hill Encyclopedia of Science and Technology* (New York, McGraw-Hill, 1960); and David Teece, *The Multinational Corporation and the Resource Cost of International Technology Transfer* (Cambridge: Ballinger, 1976). Many of the concepts and results contained in this chapter are discussed in more detail in the latter publication.

8. *Ibid.*

9. An innovation is said to have been commercialized if it has already been applied in a facility of economic size which is essentially nonexperimental in nature. Thus, pilot plant or prototype application is not considered to represent commercialization.

10. These categories are (a) process or design engineering costs and related consultation for process innovations, or (b) production engineering expenses for product innovations; and (c) costs of engineering supervision and consultation (salaries plus travel and living) for the plant engineering.

11. Drafting costs can be considered an implementation cost rather than a transfer cost, the implication being that if the host country does not have these skills, the viability

The third group of costs are those of R and D personnel (salaries and expenses) during all phases of the tranfer project. These are not the R and D costs associated with developing the underlying process or product innovations. Rather, they are the R and D costs associated with solving unexpected transfer problems and adapting or modifying the technology. For instance, research scientists may be utilized during the transfer if new and unusual technical problems are encountered[12] with the production inputs. These R and D costs are generally small or non-existent for international transfers falling into the "design transfer" category.

The fourth group of costs are the pre-start-up training costs and the "excess manufacturing costs." The latter represent the learning and debugging costs incurred during the start-up phase, and before the plant achieves the design performance specifications. It is quite possible that no marketable output will be produced during the initial phases of the start-up. Nevertheless, normal labor, materials, utilities, and depreciation costs will be incurred, together with the costs of the extra supervisory personnel that will inevitably be required to assist in the start-up. The operating losses incurred during initial production are very often a close approximation to excess manufacturing costs.[13]

---

and cost of the project is unlikely to be affected. The advisory and consulting costs, on the other hand, represent transfer costs since these activities are necessary if the technology is to be adjusted to the local circumstance and requirements. Clearly, if an existing plant was to be duplicated in its own environment, consulting costs could be expected to go to zero, whereas routine drafting would still have to be performed.

12. Referring to process technologies, it is possible that differences in feedstocks among various locations may create problems that only research scientists can effectively handle. Similarly, changes in atmospheric conditions or water supply could have unexpected consequences for some highly complex processes.

13. An important consideration is the extent to which excess manufacturing costs correctly reflect technology transfer costs rather than the costs of discovering and overcoming the idiosyncrasies of a new plant. One way to confront the issue is to consider the level of excess manufacturing costs when an absolutely identical plant is constructed in a location adjacent to an existing plant. Further, assume the second plant embodies the same technology as the first plant, and the labor force from the first is transferred to the second for the purpose of performing the manufacturing start-up. The assumption is that under these circumstances excess manufacturing costs in the second plant will be zero, or very nearly so. The correctness of this assumption was corroborated by a subsample of project managers subsequently questioned about this matter. The postulated circumstance would be identical to shutting down the first plant and then starting it up again. Some excess manufacturing costs might be incurred during the initial hours of operation if the plant embodies flow process technology. (For the projects in the sample the average duration of the manufacturing start-up was 8.2 months.) However, these costs are unlikely to be of sufficient magnitude to challenge the validity of classifying excess manufacturing costs as a component of technology transfer costs.

5. *Transfer Costs: Data and Hypotheses*

The above definition was used to calculate the transfer costs for twenty-six projects. The results are presented in Table 4.2. The costs are given in absolute dollars, and then normalized by total project costs.[14] For the sample as a whole, transfer costs average 19 percent of total project costs. Clearly, the data do not support the notion that technology is a stock of blueprints usable at nominal cost to all. Nevertheless, there is considerable variation in the sample data, with transfer costs ranging from 2 percent to 59 percent of total project costs. The number of factors influencing transfer costs is undoubtedly very great, but some factors are likely to have a more pervasive influence than others. The discussion to follow is restricted to hypotheses for which statistical testing

TABLE 4.2

*Sample Data on the Resource Costs of Technology
Transfer: Twenty-six International Projects*

CHEMICALS AND PETROLEUM REFINING		MACHINERY	
TRANSFER COSTS: DOLLAR AMOUNT (THOUSANDS)	TRANSFER COSTS: DOLLAR AMOUNT / TOTAL PROJECT COST (PERCENT)	TRANSFER COSTS: DOLLAR AMOUNT (THOUSANDS)	TRANSFER COSTS: DOLLAR AMOUNT / TOTAL PROJECT COST (PERCENT)
49	18	198	26
185	8	360	32
683	11	1,006	38
137	17	5,850	45
449	8	555	10
362	7	1,530	42
643	6	33	59
75	10	968	24
780	13	270	45
2,142	6		
161	2		
586	7		
877	7		
66	4		
2,850	19		
7,425	22		
3,341	4		

14. Total project costs are measured according to the inside boundary limits definition commonly employed by project accountants. Installations outside the plant perimeter are thereby excluded.

is feasible, given the available data. Two groupings of testable hypotheses can be identified: characteristics of the technology/transferor, and characteristics of the transferee/host country.

A critical factor in the transfer of technology is the extent to which the technology is completely understood by the transferor. The number of manufacturing start-ups[15] or applications that the transferor has already conducted with a specific technology can be used as an index of this knowledge.[16] An increase in the number of applications is likely to lower transfer costs since with each start-up additional knowledge about the technology is acquired. Since no two manufacturing start-ups are identical, each start-up provides the firm with the opportunity to observe the effects of different operating parameters and differences in equipment design. Each application can be regarded as a new experiment that yields new information and new experience.[17] Transfer will be facilitated the more fully the technology is understood. Besides these engineering economies, additional applications provide expanded opportunities for the pre-start-up training of the labor force. Clearly, if identical or similar plants exist elsewhere, then experienced operators from these plants can be used to assist the start-up in the new plant. In addition, untrained operators can be brought into existing plants for pre-start-up training.

The second variable to be considered is the age of the technology. The age of the technology is defined as the number of years elapsing from the beginning of the first commercial application of the technology[18] (anywhere in the world) to the end[19] of the technology transfer program. The age of a technology will determine the stability of the engineering designs and the transferor's knowledge of the manufacturing procedures. The older technology, *ceteris paribus,* the greater have been the opportunities for interaction between the development groups and the manufacturing and operating groups within the firm. Problems stand a better chance of already having been ironed out, and the drawings are

15. Manufacturing start-ups are synonymous with the number of applications of the technology. If a new plant is built for each application, it would also be synonymous with the number of plants that are built that utilize the technology.

16. Corporations engaged in technology transfer ventures not grounded on their own technology are known to have encountered massive transfer problems and costs.

17. The first application represents first commercialization of the technology. This will result in the creation of a set of basic engineering drawings and specifications. Duplication and alteration of these for subsequent start-ups will involve a modest cost compared to the initial cost of constructing them.

18. If there is more than one key innovation embodied in the technology, then the date of commercial application of the most recent key innovation is the reference data.

19. Age is defined up to the end of the transfer program since any knowledge about the technology acquired up to this point is potentially useful for the transfer. For the very first start-up, age will be the length of the transfer minus the development overlap.

likely to be more secure. Further, since technology is not embodied in drawings alone, there is a great deal of uncodified information—the relevant "art." This kind of knowledge exists among the supervisors, engineers, and operators. As the age of the technology increases, more individuals in the firm have the opportunity to acquire this noncodified information, and hence are potentially available to assist in the transfer. There will, however, be some point after which greater age will begin to increase the cost of transfer. When the length of stay of corporate personnel begins to be outstripped by the age of technology, then the noncodified dimensions of design knowledge may be lost to the firm.[20]

It is necessary to distinguish the cost reductions resulting from additional start-ups from the cost reductions resulting from greater age of the technology. For continuous flow technologies, additional applications of an innovation in entirely new plants will allow experimentation with scale and with the basic parameters of the design. This will generate a greater understanding of the technology. On the other hand, greater age, given the number of applications or start-ups, generally permits experimentation only with operating parameters, the design of the plant remaining fixed throughout.

The third technology variable to be considered is the number of firms utilizing the technology, or one that is "similar and competitive." This is taken to represent the degree to which the innovation and the associated manufacturing technology is already diffused throughout the industry. The greater the number of firms with the same or similar and competitive technology, then the greater the likelihood that technology is more generally available and can therefore be acquired at lower cost.[21]

These technology variables and the attendant hypotheses begin to take on some extra significance when viewed together. Taken singly, they define the technology to only a limited degree. Together, they hypothesize, *ceteris paribus*, that the most difficult and hence costly technology to transfer is characterized by very few previous applications, a short elapsed time since development, and limited diffusion. Technology displaying such characteristics can be termed "leading edge" technology. "Leading edge" technology is likely to be in a state of flux; the

20. In the limit, the firm could terminate its utilization of a particular technology, and the noncodified information associated with it could be gradually lost forever as the technology becomes historic. Further, the drawings associated with technology that is very old may suffer from so many small alterations that the very essence of even the codified technology may become quite obscure. Since none of the technology transfer projects in the sample was historic in the above sense, the relevant range of the hypothesized age-transfer cost function involves an inverse relationship between the age of the technology and the cost of transfer.

21. An identification problem may exist here because more firms may have applied the technology because the transfer cost is low.

engineering drawings will be constantly changing, thus frustrating the transfer. In comparison, state-of-the-art technology is hypothesized, *ceteris paribus,* to involve lower transfer costs since the engineering drawings are more likely to be finalized and the fundamentals of the technology stand a better chance of being more fully understood.

The technical and managerial competence of the transferee is also an important determinant of the ease with which technology can be absorbed. The years of manufacturing experience of the transferee in a given four-digit ISIC industry is used as an index of the extent to which managers, engineers, and operators have command over the general manufacturing skills of an industry. A firm skilled in the manufacture of a group of products is likely to have less difficulty absorbing a new innovation in that industry group than is the firm which has had no previous experience manufacturing products in a particular industry group. Older enterprises, with their skilled manufacturing personnel, seem more likely to be able to understand and apply codified knowledge to the manufacture of a new product, or the utilization of a new process.[22]

Another variable to be considered is the size of the transferee. Although less compelling, the reasoning behind the hypothesis that transfer costs decline with firm size is that larger firms generally have a wider spectrum of technical and managerial talent that can be called on for assistance during the transfer. A small firm may be technically and managerially quite competent yet unable to absorb new technology easily because of the extra demands placed on its scarce managerial and technical manpower. Consultants may have to be engaged by the smaller firms to perform tasks that are typically handled internally in larger firms.

A third variable considered is the R and D activity of the transferee. When unusual technical problems are unexpectedly encountered, an in-house R and D capability is likely to be of value. Oshima has argued that the R and D capability of Japanese firms facilitated the low-cost importation of foreign technology by Japanese firms. The R and D to sales ratio of the transferee is taken as an index of its R and D capability, and an inverse relationship between this and transfer cost is postulated.

22. According to Rawski, recent experience of the People's Republic of China shows that during at least some phases of industrialization, production experience may be a key determinant of the level and fungibility of industrial skills. Rawski notes that "with their skilled veteran workers and experienced technical persons, old industrial bases and old enterprises find it easier to tackle complicated technical problems than new enterprises and new industrial bases. With these advantages, it is the established centers which are best able to copy foreign equipment samples, to extract useful information from foreign technological publications, and to apply it to current domestic problem areas." T. Rawski, "Problems of Technology and Absorption in Chinese Industry," *American Economic Review* 65 (1975):386.

The final variable considered is designed to reflect the level of development of the host-country infrastructure, which is hypothesized to be a determinant of the cost of transfer. For example, the level of skill formation in the host country will influence the amount and type of training that the labor force will require. Similarly, if the new venture is to acquire its inputs domestically, the quality of the inputs available will undoubtedly influence the level of start-up costs. There are many other considerations of similar kind which could be discussed. However, the high degree of cross-sectional collinearity between indices of development makes the identification of separate effects statistically difficult. However, GNP per capita, a measure of productive capacity, can be expected to capture some of the above considerations, and it will be used in this study as an index of economic development. A negative relationship between transfer cost and GNP per capita is postulated.[23]

6. Determinants of the Cost of International Technology Transfer: Tests and Results

The basic model to be tested is

(4.1) $$C_i = f(U_i, G_i, E_i, R_i, S_i, N_i, P_i, Z_i),$$

where C_i is the transfer cost divided by the total project cost for the ith transfer; U_i is the number of previous applications or start-ups that the technology of the ith transfer has undergone by the transferor;[24] G_i is the age of the technology in years; E_i is the number of years of manufacturing experience that the recipient of the ith transfer has accumulated; R_i is the ratio of research and development to sales for the recipient of the ith transfer (calculated for the year the transfer commenced); S_i is the volume of sales (measured in millions of dollars) of the recipient of the ith transfer; N_i is the number of firms identified by the transferor as having a technology that is identical or "technically similar and economically competitive" to the technology underlying the ith transfer; P_i is the level of GNP per capita of the host country; and Z_i is the random error term for the ith transfer.

Since one of the best tests of any hypothesis is to look for the convergence of independent lines of evidence, the testing of this model will

23. The sample did not include countries where high GNP statistics were grossly dependent on oil revenues.

24. The number of previous manufacturing start-ups was significant in this phase of the work only when it was included as a dummy variable taking the value 1 if there had been no previous manufacturing start-ups of this technology by the transferring firm and zero otherwise.

proceed in two phases. First, cross-section data on twenty-six completed projects is utilized in a linear version of the model estimated by ordinary least-squares procedures. Second, cost estimates provided by project managers for comparable projects are pooled to test a more specific non-linear version of the model.

The model to be tested in the first phase of our statistical work is

$$(4.2) \quad C_i = \alpha_0 + \alpha_1 \bar{U}_i + \alpha_2 G_i + \alpha_3 E_i + \alpha_4 R_i + \alpha_5 S_i + \alpha_6 N_i + \alpha_7 P_i + Z_i,$$

where $\bar{U}_i$ is a dummy variable taking the value 1 if the ith transfer represents the first manufacturing start-up, and zero otherwise. $\bar{U}_i$ is used rather than U_i for empirical reasons since the first start-up is often of critical importance. The sample was dichotomized because of the large differences between continuous flow process technology and product technology. One category includes chemicals and petroleum refining, and the other includes machinery (see Table 4.1).

The results in Table 4.3 indicate that in chemicals and petroleum refining $\bar{U}_i$ and N_i are significant at the 0.05 level and have the expected signs. In the machinery category the variables N_i, G_i, and E_i all have the expected signs and are significant at the 0.05 level. Since N_i and E_i are significant (or close to it) in both industry groupings, the evidence seems to support quite strongly the hypothesis that transfer costs decline as the number of firms with identical or "similar and competitive" technology increases, and as the experience of the transferee increases. However, R_i and P_i were not significant in any of the equations, and although S_i has the expected sign and approaches significance in one of the regressions, it is not statistically significant.[25]

Thus, the results support (at least to some extent) a number of hypotheses advanced earlier, but there are differences in the size of coefficients as well as in the specification of the equations between the industry groups. In particular, the novelty variable $\bar{U}_i$ is significant in chemicals and petroleum refining, but insignificant in machinery. The converse is true for the age variable G_i. The reason may be that there

---

25. Multicollinearity does not appear to be a serious problem in any of the equations. Correlations between pairs of the independent variables were never significant at the 0.05 level. The stability of the regression coefficients further suggests that multicollinearity is not a serious problem. Dummy variables were introduced to test for the effects of the organizational relationship between transferor and transferee (affiliate/nonaffiliate, public enterprise/private enterprise), but they were not found to be a statistically significant determinant of transfer costs. Application of a forward step-by-step procedure did not reveal a preferred subset of variables. However, it is possible that the correct model is the simultaneous equation model $C_i = f(N_i, \ldots), N_i = f(C_i, \ldots)$. To eliminate simultaneous equation bias it would be desirable to use a two-stage procedure. However, no attempt along this line was made in this exploratory study.

TABLE 4.3

*Regression Coefficients and t Statistics in Regression Equations
to Explain C (the Cost of Transfer)*

| INDEPENDENT VARIABLE | CHEMICALS AND PETROLEUM REFINING | | MACHINERY | |
	EQUATION (1)*	EQUATION (2)*	EQUATION (1)*	EQUATION (2)*
Constant	12.79 (6.82)	13.42 (6.98)	16.67 (8.27)	65.98 (6.60)
Novelty dummy variable $\bar{U}$†	6.73 (1.92)	6.11 (1.75)	—	1.62 (0.15)
Number of firms	−0.37 (−2.06)	−0.39 (−2.22)	−1.29 (−2.28)	−1.26 (−1.95)
Age of technology (years)	—	—	−2.43 (−3.53)	−2.35 (−2.51)
Experience of transferee (years in four-digit ISIC)	−0.09 (−1.66)	−0.08 (−1.42)	−0.84 (−3.37)	−0.85 (−2.95)
Size of transferee (millions of dollars of sales)	—	−0.0009 (−1.18)	—	—
Number of observations	17	17	9	9
R^2	0.56	0.61	0.78	0.78
F	5.66	4.73	6.00	3.22
Significance level of F	0.01	0.02	0.04	0.12

*Omitted coefficient indicates variable dropped from the regression equation.
†Footnote 24.

exists relatively less latitude for production experimentation with continuous flow process technology than with product technology. Once the plant is constructed, the extent to which the design parameters can be changed is rather minimal because of the degree of interdependence in the production system. In comparison, many product technologies allow greater design flexibility. Innumerable small changes to the technology are very often possible without massive reconstruction of the plant. It is also of interest that the coefficient of the experience variable E_i is considerably larger in machinery than in chemicals and petroleum refining. This is consistent with other findings that reveal important learning economies in fabrication and assembling.

The above analysis is handicapped by the small sample size and the very high costs of adding additional observations. Limited variation in exogenous variables coupled with the problem of omitted variables can imply difficulties with bias and identification. Thus, we carried out a second phase of the statistical work. For the projects in the sample, a procedure was devised to hold the missing variables constant while generating large variation in the exogenous variables. The respondent firms were asked to estimate how the total transfer costs would vary for each project if one particular exogenous variable happened to take a different value, assuming all other variables remain constant. The responses were included only if the exercise generated circumstances within the bounds of an executive's experience. Given these limits, the change specified was large so as to increase the precision of the estimated effects. Generally the actual value of a selected variable was hypothesized first to halve and then to double. The estimated impact on transfer costs was noted. The exercise was performed for the following independent variables: the number of applications or start-ups that the technology has undergone; the age of technology; the number of years of previous manufacturing experience possessed by the transferee in a given four-digit industry; the research and development expenditures to sales ratio for the transferee; the size (measured by sales value) of the transferee. For each variable this exercise generated at most three observations (including the actual) for each project. Pooling across projects produces enough observations for ordinary least-squares regression analysis.

The estimation procedure is commenced by assuming that the shape of the cost function can be represented by the following relatively simple but quite specific equation:

(4.3)
$$C_j = V e^{\frac{\phi}{X_j}}.$$

C is the estimated transfer cost as a percentage of total project cost, X represents the value of various independent variables, j refers to the jth observation.

With this specification, the transfer cost for a project asymptotically approaches a minimum nonzero value as the value of each X increases. That is, as X goes to infinity, C goes to V. Therefore, V is the minimum transfer cost with respect to the X variable. However, there is no maximum cost asymptote for the range of the data. The expression for the elasticity of transfer cost with respect to X is given by

(4.4)
$$\frac{-X \, dC}{C \, dX} = \frac{\phi}{X}.$$

Thus, for a specified value of X, the elasticity of transfer cost with respect to X is determined by ϕ. Hence, the elasticity depends only on ϕ and X. In order to estimate the function, the log of the arguments in (4.3) is taken:

(4.5)
$$\log C_j = \log V + \frac{\phi}{X_j}.$$

Dummy variables are used to pool the observations across projects. Inclusion of dummy variables allows the minimum cost asymptote to vary from project to project. It is assumed that ϕ is constant across projects. These assumptions provide a pooled sample with intercepts that vary across projects.

Ordinary least-squares regressions of $\log C_j$ on the dummy variables and $1/X_j$ then proceeded for five different X variables, and for five data sets. These were: total transfers; transfers within the chemical and petroleum refining category; transfers in the machinery category; transfers of chemicals and petroleum refining technology to developed countries; and transfers of chemical and petroleum refining technology to less developed countries.[26] The Chow test of equality between sets of coefficients in two linear regressions revealed that the separation of the sample along industry lines was valid, except for the research and development variable. However, there was no statistically valid reason for disaggregating the chemical and petroleum refining subsample according to differences in GNP per capita in the host countries.

The results of the estimation are contained in Table 4.4. The high R^2 values are partly because the large across-project variation in costs is being captured by the project dummies. The intercept term was always

TABLE 4.4

Estimated Values of ϕ
(Obtained from Regressing $\log C_j$ on ϕ/X_j)
Together with Corresponding t-Statistics, Sample Size,
Degrees of Freedom, and Coefficient of Determination

VARIABLE	ϕ	t-STATISTIC	SAMPLE SIZE	DEGREES OF FREEDOM	R^2
Start-ups					
Chemicals and petroleum refining	0.46	4.23	45	25	0.92
Machinery	0.19	1.76	20	10	0.91
Age					
Chemicals and petroleum refining	0.04	1.29	47	30	0.89
Machinery	0.41	2.19	21	13	0.94
Experience					
Chemicals and petroleum refining	0.007	0.85	52	33	0.78
Machinery	0.57	6.08	23	14	0.91
Size					
Chemicals and petroleum refining	0.008	1.17	54	35	0.88
Machinery	0.081	5.18	17	10	0.99
R and D ÷ sales					
Total sample	0.06	1.58	59	30	0.90

26. A purely arbitrary classification was used where less developed countries were defined as those with GNP per capita less than $1,000. Note too that, if the errors are not independent in equation (4.5), the t-statistics may be overstated.

highly significant and the coefficients on all the dummies were significantly different from each other. All of the coefficients are significantly greater than zero at the 0.20 level, and the age of the technology, the number of manufacturing start-ups, transferee size and experience achieve at least the 0.05 significance level in one or other of the subsamples. In several cases the coefficients are highly significant, providing strong statistical support for the hypotheses that have been advanced. The number of previous applications once again has a sizable impact. Age and manufacturing experience are particularly important in the machinery category.

The calculation of elasticities allows interpretation and comparisons of estimated effects. Average or point elasticities for some typical sample values of X are presented in Table 4.5. These estimates suggest that in the chemicals and petroleum-refining category, the second start-up could lower transfer costs by 34 percent below the first start-up, other variables held constant. The corresponding change for conducting a third start-up is 19 percent. The other elasticities can be interpreted similarly.

7. Differences between International and Domestic Technology Transfer

Although this is primarily a study of international technology transfer, it is apparent that many of the characteristics of international technology transfer are also characteristic of the technology transfer that occurs within national borders. Nonetheless, there are differences. For instance, distance and communication costs very often differentiate international from domestic transfers. Although the communications revolution of the twentieth century has enormously reduced the barriers imposed by distance,[27] the costs of international communication are often significant.[28] Language differences can also add to communication costs, especially if the translation of engineering drawings is required. The experience of Polyspinners Ltd. at Mogilev in the Soviet Union is ample testimony to the extra costs that can be encountered.[29] International differences in units of measurements and engineering standards can com-

27. Facsimile equipment exists that can be used to transmit messages and drawings across the Atlantic instantaneously.

28. One of the participating companies indicated that travel, telegraph, freight, and insurance added about 10 percent to the total cost of a project established in New Zealand.

29. The project manager estimated that documentation alone cost £500,000, and the translation a similar amount. See D. Jones, "The 'Extra Costs' in Europe's Biggest Synthetic Fiber Complex at Mogilev, U.S.S.R.," *Worldwide Projects and Installations* (1973).

TABLE 4.5

*Arc or Point Elasticity of Transfer Costs
with Respect to Number of Start-ups, Age of Technology,
Experience, Size and R and D / Sales of Transferee*

INDEPENDENT VARIABLE	CHEMICALS AND PETROLEUM REFINING	MACHINERY
	Arc elasticity	
Number of start-ups		
1–2	0.34	0.14
2–3	0.19	0.08
3–4	0.13	0.05
9–10	0.05	0.02
14–15	0.03	0.01
Age of technology	Point elasticity	
(years)		
1	0.04	0.41
2	0.02	0.20
3	0.01	0.14
10	0.00	0.04
20	0.00	0.02
Experience of	Point elasticity	
transferee (years)		
1	0.007	0.57
2	0.003	0.28
3	0.002	0.19
10	0.001	0.06
20	0.000	0.03
Size of transferee	Point elasticity	
(millions of sales dollars)		
1	0.008	0.081
10	0.001	0.008
20	0.000	0.004
100	0.000	0.001
1,000	0.000	0.000
R and D / Sales of transferee	Total sample	
(%)	point elasticity	
1	0.06	
2	0.03	
3	0.02	
4	0.01	
5	0.01	
6	0.01	

pound the problems encountered.[30] Additional sources of difficulty are rooted in the cultural and attitudinal differences among nations, as well as differences in the level of economic development and the attendant socioeconomic structure.

It is of interest to know the magnitude and determinants of the "international component" of the transfer cost. Unfortunately, foreign and domestic transfers are rarely identical in scope or in timing, and so it is not possible to gather comparative data on implemented projects at home and abroad. It was therefore found necessary to rely on estimates provided by the firms involved in international transfers. For the projects in the sample, project managers were asked to estimate the dollar amount by which transfer costs would be different if the international transfers in the sample had occurred domestically, holding firm and technology characteristics constant. The procedure was designed to highlight the effects of country characteristics such as differences in language, differences in engineering and measurement standards, differences in economic infrastructure and business environment, and geographical distance from the transferor. The international component of the transfer cost for the projects in the sample could be obtained by subtracting the estimated transfer cost from the actual transfer cost. The data, contained in Table 4.6, reveal that the difference in cost is not always positive. This indicates that in at least some of the cases, the international transfer of an innovation was estimated to cost less than a comparable domestic transfer. This may seem paradoxical at first, given that international technology transfer generally augments the transfer activities that have to be performed.[31] An analysis of the determinants of the international component of transfer costs may yield an explanation.

Several hypotheses are presented and tested. The first is that the difference is large and positive when the technology has not been previously commercialized. National boundaries are often surrogates for cultural and language barriers, differences in methods and standards of measurement, and distance from the home country. During first commercialization of a product or process, there are generally enormous information flows across the development-manufacturing interface. The hypothesis is that placing a national boundary at this interface can complicate matters considerably, and escalate the costs enormously. The second hypothesis is that transfers to government enterprises in centrally planned economies will involve higher transfer costs. Transferors can

30. See J. Meursinge, "Practical Experience in the Transfer of Technology," *Technology and Culture* 12 (1971):469–70.

31. The source of the apparent paradox may be differences in labor costs. Nevertheless, the identification of the characteristics of transfers for which international transfer costs less than domestic transfer is a topic of importance.

TABLE 4.6

International Component* of Transfer Cost

CHEMICALS AND PETROLEUM REFINING		MACHINERY	
DOLLAR AMOUNT (THOUSANDS)	AS PERCENT OF ACTUAL TRANSFER COST	DOLLAR AMOUNT (THOUSANDS)	AS PERCENT OF ACTUAL TRANSFER COST
3.03	6.07	35.55	17.88
0.00	0.00	−399.37	−110.93
−12.81	−1.87	50.06	4.93
43.90	31.00	830.70	14.20
0.00	0.00	−4.59	−0.02
5.17	1.42	226.80	14.82
132.75	20.63	0.67	1.99
0.00	0.00	−134.40	−13.87
342.00	43.84	34.98	12.95
0.00	—		
0.00	—		
0.00	0.00		
−10.77	−6.66		
−50.16	−8.52		
0.00	—		
637.32	72.60		
−1.33	−1.99		
1,723.81	60.48		
1,370.25	18.45		
524.25	15.69		

*Amount of actual transfer costs attributable to the fact that transfer was international rather than domestic. (Accordingly negative values indicate that firms estimated that transfer costs would be higher had the transfer been domestic.) In general, these numbers were derived by taking the weighted average of estimated changes in the various identifiable components of transfer costs.

expect numerous delays and large documentation requirements. The third hypothesis is that the less the diffusion of the technology, measured as before by the number of firms utilizing the innovation, the greater the positive differential associated with international technology transfer. The fourth hypothesis is that whereas, in general, low levels of economic development are likely to add to transfer costs because of inadequacies in the economic infrastructure, this may be more than offset, in some circumstances, by low labor costs. Labor costs can have a substantial impact on excess manufacturing costs, especially in relatively labor-intensive industries. Since machinery manufacture is relatively labor-intensive, the hypothesis is that the GNP per capita in the host country is positively related to the transfer cost differential in this classification, but is negatively associated with the differential in the chemicals and petroleum refining category.

To test these hypotheses it is assumed that

$$D_i = \alpha_0 + \alpha_1 d_i + \alpha_2 \bar{U}_i + \alpha_3 N_i + \alpha_4 P_i + Z_i \, ,$$

where D_i is the "international component" as a percentage of actual transfer cost for the ith transfer. d_i is a dummy variable which takes the value 1 if the recipient of the ith transfer is a government enterprise in a centrally planned economy, and zero otherwise. The other variables carry the same definitions as previously. Least-squares estimates of the α's were obtained, the results being

Chemicals and petroleum:

$$D_i = 0.285 + 3.84d_i + 4.46\bar{U}_i \quad (n = 17, r^2 = 0.71).$$
$$(0.91) \quad (5.01) \quad (4.89)$$

Machinery:[32]

$$D_i = -8.59 - 1.39N_i + 0.005P_i \quad (n = 9, r^2 = 0.94).$$
$$(-1.96) \quad (-5.98) \quad (3.90)$$

The hypotheses are to some extent borne out by the data, but the small sample size must counsel great caution in the interpretation of these results.[33] In chemicals and petroleum, the results indicate that transfers to government enterprises, and transfers before first commercialization, involve substantial extra costs. In machinery, both N_i and P_i are significant (despite the small number of observations), although they are insignificant in chemicals and petroleum refining (where there are more than twice as many degrees of freedom). Taken at face value, there seems to be no evidence that the level of host-country development and the degree of diffusion of an innovation affect the international-domestic transfer cost differential in the chemicals and petroleum grouping. This calls for an explanation. The diffusion variable N_i is taken to indicate the degree to which the requisite skills are generally available. The statistical results suggest that the relevant skills for highly capital-intensive industries, such as chemicals and petroleum refining, are more easily

32. d_i was omitted from the machinery regression since none of the actual transfers in this category was to government enterprises in centrally planned economies.

33. If the second observation in the machinery category in Table 4.6 is excluded, and the regression results recomputed, the estimates of the coefficients exhibit considerable instability and the "goodness of fit" deteriorates. The estimated equation is

$$D_i = -4.96 - 0.66N_i + 0.003P_i \quad (n = 8, r^2 = 0.45).$$
$$(1.14) \quad (2.40) \quad (1.94)$$

These estimates are nevertheless significant at the 0.05 level with a one-tail test.

transferred internationally than are the requisite skills in the machinery category.[34] Furthermore, P_i was not significant in chemicals and petroleum refining, suggesting that costs of transfer are independent of the level of economic development in this category. This is consistent with speculation that international transfer is no more difficult than domestic transfer when the underlying technology is highly capital-intensive.

8. Summary

Technology transfer costs are defined here as the costs of transmitting and absorbing all of the relevant unembodied knowledge. They include the cost of pre-engineering technological exchanges during which the basic characteristics of the technology are revealed to the transferee. They include the engineering costs associated with transferring the process design and the associated process engineering in the case of process innovations, or the product design and production engineering in the case of product innovations. They also include the R and D costs involved in adopting and modifying technology. And they include the pre-start-up training costs and the learning and debugging costs incurred during the start-up phase, and before the plant achieves the design performance specifications.

In our sample of international transfers of technology, transfer costs averaged about 20 percent of total project costs. The size of these transfer costs would be expected to depend on the number of manufacturing start-ups or applications that the transferor has already conducted with this technology, the age of the technology, the number of firms utilizing this technology, the number of years of manufacturing experience of the transferee, the size of the transferee, and the extent of the R and D activity of the transferee. Also, the level of development of the host country's infrastructure would be expected to influence the size of the transfer costs.

Based on our econometric results, the number of manufacturing start-ups, the age of the technology, the number of firms utilizing the technology, and the number of years of manufacturing experience of the transferee turn out to have a statistically significant effect on transfer costs. But this generally is not true of the size of transferee, the extent of its R and D activities, or the level of gross national product per capita in the country where the transferee is located. In the chemicals and

34. This is consonant with the views expressed by several project managers in the chemical industry. It was asserted that technology could be transferred with equal facility to almost anywhere in the world, including less developed countries, assuming host-government interference is held constant.

petroleum refining industries, the second start-up could lower transfer costs by about 34 percent below the first start-up, and the third start-up could lower transfer cost by about 19 percent. In the machinery industry, the manufacturing experience of the transferee and the age of the technology have very large effects on transfer costs.

Many of the characteristics of international technology transfer are also characteristic of the technology transfer that occurs within national borders. Thus, in most cases in our sample, a large proportion of the transfer costs would have been incurred even if the technology had been transferred within the United States. However, because of such factors as language barriers, internatonal differences in units of measurement and engineering standards, and cultural, attitudinal, and political differences, there generally are extra costs associated with transferring technology across national boundaries. In the chemical and petroleum refining industries, these extra costs seem to be particularly great when the technology has not been commercialized yet and when it is transferred to government enterprises in centrally planned economies. In the machinery industry, these extra costs seem to be particularly great when the technology has spread to very few firms and when the per capita income (and the prevailing wage rate) of the country in which the transferee is located is high.

5 INTERNATIONAL TECHNOLOGY TRANSFER: TIME-COST TRADEOFFS AND OVERSEAS R AND D EXPENDITURES

1. Introduction

The previous chapter dealt with the cost of transferring technology. In this chapter, we recognize that the cost of designing, constructing, and starting up a manufacturing project abroad based on U.S. technology depends on the time span between project commencement and project completion. In other words, there is a time-cost tradeoff. In previous publications, we have studied the time-cost tradeoffs for the innovation process in the United States.[1] Now we investigate such tradeoffs for the process of establishing a plant abroad based on U.S. technology.

Besides establishing plants abroad, U.S.-based firms have also established R and D facilities there. In recent years, the overseas R and D activities of U.S.-based firms have become the focus of controversy for reasons given below. Unfortunately, however, economists have devoted little or no attention to even the most basic questions concerning these activities, with the result that little is known about their nature, extent, and rationale. Another purpose of this chapter is to help fill this gap.

2. Foundations of the Time-Cost Tradeoff

While the existence of a time-cost tradeoff for research and development has been recognized and demonstrated, the applicability of this

---

1. In particular, see Mansfield, Rapoport, Schnee, Wagner, and Hamburger, *op. cit.* (cited in Chapter 1, footnote 6). The importance of such tradeoffs has been recognized in a number of different applications, and critical path analysis has been used to assist in the efficient scheduling of complex development and construction projects.

discovery to nonresearch activity has at the same time been questioned.[2] Foreign direct investment, not obviously a research-related activity, nevertheless seems to be rooted in the exploitation of technological know-how. Furthermore, the application of U.S. technology abroad is often replete with technological uncertainty, just as is the development of new technology itself. The utilization of a technological innovation in a new context is likely to require, among other things, adjustment of some of the basic design parameters. For example, differences in the market size between home and abroad will induce scale adaptations to the plant; and differences in materials inputs, operator skills and engineering standards will frequently necessitate design changes in the process and / or the product. The implementation of design changes will produce uncertain responses in the quality and cost of the final product. Although the uncertainties generated are undoubtedly modest compared to those encountered during the original product or process development, they are still important. When uncertainty precludes immediate identification of the best design, it may be desirable to "hedge" by supporting several different designs. By incurring higher project costs, hedging can reduce the project time relative to a procedure which explores different designs sequentially.

Besides "hedging" activities there are a number of other procedures that can be used to reduce project time, but they can all be expected to increase project costs. As additional engineers are brought on to the project to speed the design, diminishing returns can generally be expected. The concomitant increases in job segmentation will eventually augment coordination costs. Attempts to reduce project time by speeding equipment procurement can also be expected to increase project costs. Lead-times on major items of equipment can be reduced in a number of ways. For instance, the multinational firm can bypass the equipment bidding procedure and the attendant delays by negotiating cost-plus contracts with equipment suppliers. The disabilities and costs associated with this kind of contracting have been set out adequately elsewhere. An alternative procedure is to solicit bids before the plant and equipment designs have been finalized. This may save several weeks, but firms generally run the risk of incurring penalty fees if the design specifications are subsequently modified. A number of procedures are also available to reduce manufacturing start-up time, which commonly accounts for about 20 percent of total project time. For instance, the number and duration of pre-start-up training programs can be increased. A more

2. Some people argue that the time-cost tradeoff in research and development has "few close parallels in nonresearch activities." See F. M. Scherer, "Government Research and Development Programs," in *Measuring Benefits of Government Expenditures*, ed. R. Dorfman (Washington, D.C.: The Brookings Institution, 1965).

FIGURE 5.1

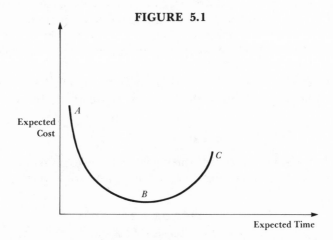

radical and costly procedure to facilitate a quick and smooth start-up would involve the importation of large numbers of trained operators from established plants to assist during the start-up period. Of course, if the new plant is the first of its kind, there may be little advantage to be gained from such costly procedures.

All of these various considerations provide the foundation for pos-tulating a time-cost tradeoff that within some range has a negative slope and is convex to the origin. If some costs are proportional to the project's duration, increasing project time need not always lower expected costs. The postulated tradeoff function is U-shaped (see Figure 5.1). Clearly, the firm will not wish to operate to the right of B under any sort of sensible conditions, and so the range of the tradeoff function that is of greatest interest is AB.

3. Estimation of the Time-Cost Tradeoff

The relevant tradeoff function is hypothesized to be negatively sloped and convex to the origin, $\partial C / \partial T < 0$, $\partial^2 C / \partial T^2 > 0$, where C is the expected project cost, and T is the expected project time. In order to test this hypothesis, data were obtained for a sample of twenty international projects. All of the projects embodied U.S. technology, and fifteen of the twenty were in chemicals or petroleum refining. The remainder were in the machinery industry. The projects varied considerably in size and in geographical dispersion.

Project managers were asked to estimate the percentage change in the actual cost of the project that would result from expected changes in the actual time.[3] The actual project cost and time were used as reference points. The project managers were also asked to estimate the minimum possible time in which the project could be completed. Five observations on time and cost were obtained for each project, including the actual time and cost.[4]

In order to estimate the function, the assumption is made that the downward sloping section of the time-cost tradeoff can be represented by

(5.1) $$C = V \exp\{\phi / ((t / \alpha) - 1)\},$$

where C is the expected cost of the project, t is the expected time, and V, α, and ϕ are parameters that vary from project to project. Figure 5.2

FIGURE 5.2
Graphical Representation of Equation 5.1

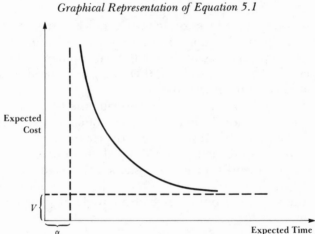

3. Managers were asked to estimate the change in cost that would result for different time spans, such as half the actual time, twice the actual time, actual time ± 10 percent, and so forth. Generally, five points on the negatively sloped portion of the tradeoff were obtained, including the actual cost-time configuration. Since this was an experiment conducted by a disinterested party, and since the confidentiality of the data was guaranteed, the respondents had no incentive to distort the data deliberately. Furthermore, the time-cost tradeoff by itself implies very little about the optimal scheduling of the project. To discover this, a revenue function must also be estimated. The "optimal" project time occurs when the marginal revenue from time shaving equals the marginal cost from time shaving. Accordingly, the data supplied could not possibly be used to make references about the performance records of the respondent managers. These factors attenuate whatever opportunistic proclivities managers might possibly entertain.

4. Although it was decided to estimate only the negatively sloped portion of the time-cost tradeoff, it is of interest to note that for thirteen of the projects in the sample, costs would have increased if the expected time were doubled. Several respondents pointed out that inept management could quite easily create situations where it might be realized *ex post* that a project had proceeded on the positively sloped portion of the tradeoff.

shows the nature of this function. It is convex and has time and cost asymptotes.

Since C approaches V as t becomes larger, V can be considered the minimum expected cost of the project. Since t approaches α as C becomes larger, α can be considered the minimum expected time to complete the innovation. The elasticity of cost with respect to time, $(-dC/dt) \cdot (t/C)$, is equal to $\phi(t/\alpha)/[(t/\alpha) - 1]^2$. Thus, for a given value of t/α, the elasticity of cost with respect to time is determined by ϕ. A logarithmic transformation of equation (5.1) together with the addition of an error term yields

$$(5.2) \qquad \ln C_i = \ln V_i + \phi[(t_i/\alpha) - 1]^{-1} + Z_i.$$

Since for each project direct estimates of α have been obtained from the respondents, estimates of V_i and ϕ could be obtained by regressing $\ln C_i$ on $1/[(t_i/\alpha) - 1]$. The error term Z_i is assumed to be distributed with mean zero and constant variance. The results are summarized in Table 5.1. In general, it can be said that the goodness of fit is acceptable, although in each case the number of observations is very small.

TABLE 5.1

Estimates of V, α, and ϕ:
Twenty International Projects

PROJECT	V (THOUSANDS OF DOLLARS)	α (MONTHS)	ϕ	r^2
1	260	9	0.024	0.61
2	1,998	20	0.068	0.69
3	3,964	14	0.065	0.99
4	796	11	0.146	0.99
5	578	32	0.174	0.90
6	1,808	28	0.070	0.98
7	9,228	24	0.089	0.55
8	3,197	15	0.030	0.95
9	111	3	0.279	0.96
10	459	10	0.072	0.94
11	1,615	21	0.007	0.82
12	11,395	30	0.119	0.96
13	29,971	61	0.028	0.98
14	2,470	20	0.115	0.95
15	654	12	0.053	0.94
16	3,901	22	0.122	0.91
17	12,100	27	0.560	0.79
18	4,745	36	0.185	0.78
19	10,872	36	0.021	0.97
20	620	17	0.041	0.97

Using the estimated value of ϕ, the elasticity of cost with respect to time was calculated for a given value of t_i / α. (t_i was set at its actual value.) The results, given in Table 5.2, show that in fifteen of twenty cases, strategies aimed at shaving the actual time by 1 percent would raise costs by more than 1 percent. By comparison, Mansfield *et al.* discovered that for innovation, a 1-percent shaving in project time could be obtained for a

TABLE 5.2

Estimates of Elasticity of Cost with Respect to Time:
Twenty International Projects

POINT ELASTICITY	REALIZED t / α				
	1.00–1.25	1.26–1.50	1.51–1.75	1.76–2.00	TOTAL
0–0.50	1	0	1	0	2
0.51–1.00	2	1	0	0	3
1.01–1.50	1	1	0	1	3
1.51–2.00	3	0	0	0	3
2.01–2.50	2	1	0	0	3
2.51–3.00	1	1	0	0	2
3.01–3.50	2	1	0	0	3
Over 3.50	1	0	0	0	1
Totals	13	5	1	1	20

TABLE 5.3

Number of Innovations with Indicated Values
of Elasticity of Cost with Respect to Time,
at Actually Realized Values of t / α,
for Twenty-nine Innovations

VALUE OF ELASTICITY*	ACTUALLY REALIZED VALUE OF t / α			
	1.00–1.50	1.51–2.00	OVER 2.00	TOTAL
	(Numbers of Innovations)			
0	0	0	1	1
0.01–0.49	1	2	7	10
0.50–0.99	2	3	0	5
1.00–2.00	1	3	2	6
Over 2.00	5	1	1	7
Totals	9	9	11	29

Source: Mansfield *et al., op. cit.*
*Arc elasticities were computed between the point at the actually realized value of t and the point at the next lower value of t given by the respondent. The difference in t/α is often quite large.

cost increase of less than 1 percent for most of the innovation projects examined. These results are contained in Table 5.3. Thus based on these data, the elasticity of cost with respect to time for international transfer of manufacturing technology seems greater than for innovation. But the relatively small sample sizes should be borne in mind.

4. Determinants of the Elasticity of Cost with Respect to Time

It seems likely that the elasticity is related to the structure of the network of tasks involved in the technology transfer project. Unfortunately, complete data regarding the characteristics of this network for each project are not readily available. In the analysis of the determinants of the time-cost elasticity that follows, attention is confined to variables for which data are readily available. Nevertheless, it is recognized that some of the variables selected may simply be surrogates for more fundamental considerations pertaining to this network and the critical path.

The first hypothesis to be advanced is that the elasticity will be lower the longer the duration of the preliminary planning stage, hereafter stage A, relative to the other stages. Stage A usually can be telescoped, if the need arises, since it utilizes resources entirely at the firm's own command. The converse of this hypothesis is that the engineering, construction, installation, and start-up can be telescoped, but only at a relatively greater expense. This is because these activities generally involve contractual relations with other firms in which the authority of the transferor is diminished.

The second hypothesis is that the elasticity will be lower if the technology to be embodied in the new facility has been applied previously. If there has been at least one previous application, then attention can be directed away from problems of the technology per se to problems of the transfer; that is, because of the uncertainty involved, a strategy to speed up the project by maximizing overlap will present colossal problems if the technology has not been applied previously.

The third hypothesis is that the elasticity will also be determined by the size of the primary transfer agent. The primary transfer agent is defined as the enterprise with the controlling equity in the new venture. Thus, the transferor will be the primary agent if the transfer is to a subsidiary, but the transferee will be the primary agent if the transfer is to an independent enterprise. The primary agent is generally the entity that will hire the engineering contractors and authorize the bidding on key pieces of equipment. It will have the responsibility for expediting the project. It seems reasonable to hypothesize that insofar as speeding up a project requires a certain flexibility of approach, a larger organization may be handicapped by inertia and a more complex decision-making procedure. Even reaching agreement on how to proceed will take longer if more people and a longer chain of command are involved.

The fourth hypothesis is that the elasticity may be influenced by total project costs. On a priori grounds, it seems reasonable to hypothesize that because large projects require more coordination and integration of different tasks, they may be more costly to speed up than smaller projects.

Finally, it is hypothesized that the actually realized value of the elasticity is also a function of whether or not the foreign market can be satisfactorily supplied by exports in the interim. If trade barriers are not prohibitive, and if exporting has traditionally been used to supply a particular foreign market, then the marginal profits from bringing the new plant on stream more quickly are not likely to be enormous. If, on the other hand, prohibitive tariffs or import controls prohibit sourcing from abroad, then the returns from time shaving are likely to be enhanced and the elasticity measure higher since the project will be pushed a little faster.[5]

To test these hypotheses it was assumed that

$$\epsilon_i = \alpha_0 + \alpha_1 A_i + \alpha_2 \bar{U}_i + \alpha_3 S_i + \alpha_4 C_i + \alpha_5 X_i + Z_i .$$

where ϵ_i is the estimated elasticity of cost with respect to time for the ith project calculated at the actual time taken for the transfer; A_i is the percentage of total time allocated to stage A of the ith project; $\bar{U}_i$ is a dummy variable that takes the value 1 if the ith project represents the first application of the technology, and 0 otherwise; S_i is the sales volume (in millions of dollars) of the primary transfer agent for the ith project; C_i is the total cost of the ith project (measured in thousands of dollars); X_i is a dummy variable that takes the value 1 if the foreign market that the new plant will supply was previously supplied by export by the transferor. (If it was not supplied in this manner, $X_i = 0$.) Z_i is a random error term with mean zero and constant variance. Ordinary least-square estimates of the α's yielded

5. A profit-maximizing firm will choose a location on the tradeoff function such that the marginal benefit and the marginal cost of time shaving are equalized. The actually realized elasticity measure will therefore be sensitive to both the parameters of the time-cost tradeoff as well as the parameters of the benefit function. The four previous hypotheses represent parameters entering the cost function, whereas the existence of a viable export option will be a parameter entering the benefit function. Our purpose here is not to estimate a structural equation, but a rough reduced-form equation for the elasticity. Since this is the measure generally used in studies of this type, we would like to estimate its value as best we can.

(5.3) $\epsilon_i = 2.20 - 1.82A_i + 1.20\bar{U}_i + 0.00014S_i + 0.00011C_i - 0.83X_i$
        $\quad$ (10.11) (2.02)   (2.81)   (2.15)   (1.48)   (2.39)
        $$r^2 = 0.71, n = 20.$$

All of the variables are significant and take the expected signs. Most interestingly, $\bar{U}_i$ is highly significant and has a sizable impact. If the technology has not been previously commercialized, the elasticity measure is increased by 1.20. Technological uncertainty is clearly an important determinant of the elasticity estimates. X_i is also highly significant, suggesting that the incentive to shave time is reduced if the market can be simultaneously supplied from the parent plant or from foreign subsidiaries. The coefficient of S_i indicates that a billion-dollar increase in the size of the primary transfer agent increases the elasticity by 0.14. A million-dollar increase in project size increases the elasticity by a similar amount. The large coefficient of A_i indicates considerable potential for shaving time when the preliminary planning stage has been protracted. By juxtaposing these results against those derived from analysis (in our previous publications) of the time-cost tradeoff in technological innovation, it becomes apparent that variables indicating how well a technology is understood[6] are particularly significant determinants of the time-cost elasticity. Measures of project size and firm size are likewise significant in both studies.

5. Implications of the Findings

The examination and estimation of time-cost tradeoffs for the establishment of foreign manufacturing facilities have been instructive for several reasons. First, the existence of a negatively sloped function has once more been demonstrated. This is of some importance since the existence of time-cost tradeoffs has at times been treated with considerable skepticism.[7] Second, the elasticity estimates were generally greater than one, indicating that time shaving would involve rather high incremental costs. Third, it was observed that the elasticity estimates were highest for projects where the technology had not been previously commercialized, for projects that were large, and for projects carried out by the larger firms.

Some important implications follow from the analysis. First, when

6. The variables referred to are, first, the extent of the "state of the art" advance in the study of innovation, and second, a dummy variable used to indicate whether or not the technology has been previously applied. See Mansfield et al., op. cit.

7. The reference is to some military officers who have argued that although hastening a project will undoubtedly increase costs per time period, the reduction in total project time will offset the higher rates of expenditures. See Scherer, op. cit.

the transferred technology involves a change in the state of the art, the extra costs of speeding a project would seem to be considerable. The sensitivity of cost with respect to time indicates the importance of careful scheduling of such projects. Second, smaller firms implementing smaller projects seem to possess more versatility than larger firms implementing larger projects. Both of these implications rest comfortably with a view of technological change which recognizes the deficiencies as well as the strengths of the larger firm as an agent of technological change, and which also recognizes the high cost of technology transfer when an alteration of the state of the art is involved.

6. Overseas R and D Expenditures by U.S.-based Firms

Besides establishing productive facilities abroad U.S.-based multinational firms have also established overseas R and D laboratories. As pointed out at the beginning of this chapter, such overseas R and D activities have become the focus of controversy in recent years. Some observers view such activities with suspicion, since they regard them as a device to "export" R and D jobs, or as a channel through which American technology may be transmitted to actual or potential foreign competitors.[8] Others, particularly the governments of many developing (and some developed) countries, view them as highly desirable activities that will help to stimulate indigenous R and D in these countries. Indeed, the United Nations Group of Eminent Persons recommended that host countries require multinational corporations to contribute toward innovation of appropriate kinds and to encourage them to do such R and D in their overseas affiliates.[9] Since so little is known about the overseas R and D activities of U.S.-based firms, the remainder of this chapter summarizes some investigations designed to shed light on the size, nature, and rationale of these activities.

How big are the overseas R and D expenditures of U.S.-based firms, now and in the past? To help answer this question, we constructed a sample of fifty-five major manufacturing firms, this sample being divided into two parts. The first subsample, composed of thirty-five firms, included major U.S.-based firms in the chemical, petroleum, electrical equipment, metals and machinery, drugs, glass, food, and paper indus-

8. For discussion of this point of view, see Edward David, "Technology Export and National Goals," *Research Management* 17 (1974):12–16; and the Conference Board, *Overseas Research and Development by U.S. Multinationals, 1966–75* (New York: Conference Board, 1976).

9. United Nations, *The Impact of Multinational Corporations on Development and on International Relations* (New York: United Nations, 1974), p. 70.

TABLE 5.4

*Percentage of Company-financed R and D Expenditures
Carried Out Overseas, 1960–80: Fifty-five Firms*

	1960*	1965†	1970‡	1972†	1974	1980§
Thirty-five-firm subsample:						
Weighted mean	2	6	6	8	10	10
Unweighted mean	2	4	5	7	8	8
Standard deviation	3	7	7	8	10	8
Twenty-firm subsample:						
Weighted mean	—	—	4	—	9	14
Unweighted mean	—	—	5	—	8	11
Standard deviation	—	—	7	—	10	14

Source: see section 6.
*Data were not available for four firms in the thirty-five-firm subsample.
†Data were not available for one firm in the thirty-five-firm subsample.
‡Data were not available for one firm in the thirty-five-firm subsample and one firm in the twenty-firm subsample.
§Data were not available for nine firms in the thirty-five-firm subsample.

tries. The second subsample, composed of twenty firms, included major manufacturing firms in the southern New England and Middle Atlantic states; a variety of industries were included, such as chemicals, fabricated metals, and instruments. Table 5.4 shows the percentage of R and D done overseas by these firms during 1960–74 (for the first subsample) or 1970–74 (for the second subsample), as well as the estimated value of this percentage in 1980. In each subsample, about 10 percent of the total amount spent on R and D by these firms was carried out overseas in 1974. Based on the thirty-five-firm subsample, it appears that this percentage grew substantially during the 1960s and early 1970s.[10] Accord-

10. The Conference Board, *op. cit.*, has estimated the total overseas R and D expenditure of U.S.-based multinational firms in 1971–73. According to its estimates, overseas R and D constituted about 9 percent to 10 percent of total R and D expenditures carried out by U.S. firms during these years. This agrees quite well with our results for 1972 and 1974. The proportion of firms in our sample with no overseas R and D is somewhat lower than that reported by the Conference Board for firms of comparable size, but this may be due to different industry mix, the later year, or sampling error. The U.S. Department of Commerce has estimated the total overseas R and D expenditure of U.S. firms in 1966. See U.S. Department of Commerce, *U.S. Direct Investments Abroad, 1966,* Part II (undated). According to its figures, overseas R and D constituted about 7 percent of all R and D expenditures carried out by U.S. firms in 1966 (see Conference Board, *op. cit.*). This agrees reasonably well with our result for 1965. In 1978 the National Science Foundation published data for 1976 that indicated that overseas R and D constituted about 7 percent of all R and D expenditures by U.S. manufacturing firms. In 1980, it published data for 1978 indicating that overseas R and D constituted about 10 percent of all company-financed R and D.

ing to estimates provided by the firms in the sample, by 1980 about 12 percent of their R and D expenditures was expected to be made overseas.

Because of the importance in the innovation process of close communication and cooperation among R and D, marketing, production, and top management, Raymond Vernon and others have argued that a firm's R and D activities will tend to be centralized near its headquarters. Why, then, do these U.S.-based firms spend about 10 percent of their R and D dollars overseas? There are a variety of possible reasons, including the presence of environmental conditions abroad that cannot easily be matched at home, the desirability of doing R and D aimed at the special design needs of overseas markets, the availability and lower cost of skills and talents that are less readily available or more expensive at home, and the greater opportunity to monitor what is going on in relevant scientific and technical fields abroad. In our sample, practically all of the firms doing R and D overseas say that the principal reason is to respond to special design needs of overseas markets. In their view, there are great advantages in doing R and D of this sort in close contact with the relevant overseas markets and manufacturing units of the firm.

7. Factors Influencing the Percentage of a Firm's R and D Expenditures Carried Out Overseas

What determines the percentage of its R and D that a firm conducts overseas? Given the fact that overseas laboratories seem to be so closely geared to the special design needs of foreign markets (and the firm's overseas plants), we would expect that the percentage of a firm's R and D expenditures carried out overseas would be directly related to the percentage of the firm's sales that is derived from abroad. Firms with relatively small foreign markets would be expected to spend relatively little on overseas R and D. Further, we would expect that the percentage of a firm's R and D expenditures carried out overseas would be more closely related to the percentage of its sales from foreign subsidiaries than to its percentage of sales from exports. This is because much overseas R and D is in support of foreign manufacturing operations.[11]

11. Suppose that a firm's desired R and D expenditures in a given year equal
$$R = a_u S_u + a_f S_f + a_e S_e,$$
where S_e is its export sales during the relevant year, S_f is its sales through foreign subsidiaries, S_u is its sales from domestic plants to domestic customers, a_u is the proportion of sales to domestic customers that it wants to devote to R and D, a_f is the proportion of sales through foreign subsidiaries that it wants to devote to R and D, and a_e is the proportion of export sales that it wants to devote to R and D. If only the R and D in support of foreign

Holding constant the percentage of a firm's sales that come from abroad, we would expect that the percentage of a firm's R and D expenditures that is carried out overseas would be directly related to the firm's size. Economies of scale require that R and D laboratories be a certain minimum size if they are to be relatively efficient. If it is going to establish an overseas laboratory, the firm must have a big enough prospective market (in the area served by this laboratory) to support a laboratory of minimum economic scale. If the percentage of a firm's sales that comes from abroad is held constant, the probability that this prospective market will be of the requisite size is an increasing function of the absolute size of the firm.

Further, we would expect that, holding constant both the firm's sales and its percentage of sales coming from abroad, there would be inter-industry differences in the percentage of a firm's R and D expenditures carried out overseas. For example, we would expect this percentage to be relatively high in the pharmaceutical industry because some firms, according to industry sources at least, have moved a substantial amount of R and D abroad to avoid Food and Drug Administration regulations. Also, foreign regulations sometimes require that R and D be done locally. Because of these regulatory considerations, as well as other factors discussed below, we might expect the drug firms in our sample to carry out a relatively high percentage of R and D overseas.

In addition, we would expect that, if the firm's sales, its percentage of sales coming from abroad, and its industry are held constant, there will be differences over time in the percentage of a firm's R and D carried out overseas, owing to changes in the profitability of locating R and D overseas rather than in the United States (as well as bandwagon effects).[12] In general, during the early 1970s, we would expect these effects of time to have been positive, since cost differentials and other factors favored the expansion of overseas R and D.

To test these hypotheses, we carried out two sets of computations. First, we pooled the 1970 and 1974 data for the thirty-five-firm subsam-

---

subsidiaries is done overseas, it follows that the proportion of its R and D expenditures carried out overseas equals

$$P = \frac{a_f F}{a_u U + a_f F + a_e E},$$

where F is the proportion of its sales from foreign subsidiaries, U is the proportion of its sales to domestic customers, and E is the proportion of its sales that are exports. Under these circumstances, it can be shown that $\partial P/\partial F$ is always positive, but whether or not $\partial P/\partial E$ is positive depends on whether or not $a_u > a_e$. Of course, this model is a polar case, but it illustrates the point in the text.

12. In terms of the highly simplified model in footnote 11, the a's are a function of time. (Also, as indicated previously in the text, they are a function of the firm's size.)

ple, and regressed each firm's percentage of R and D expenditures carried out overseas on its percentage of sales from abroad, its sales, an industry dummy variable, and a time dummy variable. The results are

$$(5.4) \qquad A_{it} = -1.13 + 0.73T_t + 0.15Q_{it} + 0.004S_{it} + 16.10D_i,$$
$$\qquad\quad (0.44) \quad (0.34) \quad (1.81) \quad (3.09) \qquad (5.41)$$
$$\qquad\qquad\qquad\qquad\qquad\qquad\qquad (\bar{R}^2 = 0.50; n = 51)$$

where A_{it} is the percentage of the ith firm's R and D expenditures carried out overseas in year t (1970 or 1974), T_t is a dummy variable that equals 1 if t is 1974 and 0 if t is 1970, Q_{it} is the percentage of the ith firm's sales derived from abroad in year t, S_{it} is the ith firm's sales (in millions of dollars) in year t, and D_i is a dummy variable that equals 1 if the ith firm is in the drug industry and 0 otherwise.[13] Each regression coefficient's t-ratio is given in parentheses.

Second, we pooled the 1970 and 1974 data for the twenty-firm subsample and regressed each firm's percentage of R and D expenditures carried out overseas on the same variables as in equation (5.4), except that we split the percentage of sales from abroad into two parts—the percentage of sales from foreign subsidiaries and the percentage of sales from exports—and we redefine the industry dummy to include both chemicals and drugs, not drugs alone.[14] The results are.

$$(5.5) \quad A_{it} = 2.79 + 4.40T_t + 0.322F_{it} - 0.539E_{it} - 0.00162S_{it} + 4.52C_i,$$
$$\qquad\quad (1.16) \quad (1.91) \quad (4.47) \qquad (2.32) \qquad (1.13) \qquad (1.76)$$
$$\qquad\qquad\qquad\qquad\qquad\qquad\qquad (\bar{R}^2 = 0.54; n = 39)$$

where F_{it} is the percentage of the ith firm's sales from foreign subsidiaries in year t, E_{it} is the percentage of the ith firm's sales from exports in year t, and C_i is a dummy variable that equals 1 if the ith firm is in the drug or chemical industries and 0 otherwise.

The econometric results generally are in accord with our hypotheses. As expected, equation (5.4) shows that there is a direct and statistically significant relationship between a firm's percentage of sales derived from abroad and its percentage of R and D expenditures carried out overseas. And when sales derived from abroad are disaggregated in equation (5.5), a firm's percentage of sales from foreign subsidiaries has a highly significant positive effect on its percentage of R and D expenditures carried

13. Other industry dummies were tried in equation (5.4), but D_i was the only one that was statistically significant. The reason why n is less than 70 is that data could not be obtained concerning the percentage of sales from abroad for all firms in both years.

14. The reason for this redefinition is that none of the firms in this subsample is really an ethical drug firm. (There are several such firms in the other subsample.) The closest we could come to ethical drugs is the chemical firms, some of which do some work in the drug area (broadly defined).

out overseas, while its percentage of sales from exports has a significant negative effect, which suggests that these firms' exports may be more R-and-D–intensive than their domestic sales.[15] With regard to our hypothesis that A_{it} would be directly related to S_{it}, the results of equation (5.4) bear this out, but in equation (5.5) S_{it} does not have a statistically significant effect (and its regression coefficient has the wrong sign). As expected, most of the industry and time dummies are statistically significant.[16]

8. Overseas R and D: Minimum Economic Scale and Relative Cost

As noted above, many governments, particularly of developing countries, favor the establishment in their nations of overseas R and D laboratories by U.S.-based firms. One factor influencing the practicality of establishing a laboratory of a certain type in a particular overseas location is the extent of economies of scale in such laboratories. If the minimum economic scale for a laboratory of this type is quite large, a firm must be prepared to shift considerable R and D resources abroad if the laboratory is to be competitive.[17] Despite the fact that data concerning the minimum economic scale of R and D laboratories of various types would be of value to many kinds of microeconomic studies, practically no information is available on this score. In this section, we present estimates (obtained from twenty-seven members of the thirty-five-firm subsample) of the annual R and D expenditures for an overseas laboratory of minimum economic scale. Although these estimates should be treated with caution, they are of considerable interest, since they seem to be the first systematic evidence on this topic.[18]

15. This result concerning E_{it} would be expected if $a_e > a_u$ in footnote 11 and if the extremely simple model given there were valid. However, although it may be a reasonable approximation to regard some firms' overseas R and D as being entirely in support of foreign subsidiaries, this is far from the case in other firms.

16. The industry dummy is much larger and more highly significant in equation (5.4) than in equation (5.5) because, as pointed out in footnote 14, none of the firms in equation (5.5) is really an ethical drug firm. Because of sampling variation, the estimate of D_i in equation (5.4) is probably too large. One of the drug firms in our sample carried out an unusually large percentage of its R and D overseas.

17. By minimum economic scale, we mean the smallest scale that realizes all, or practically all, of the relevant economies of scale.

18. Christopher Freeman et al., "Research and Development in Electronic Capital Goods," National Institute Economic Review 34 (1965):40 have presented some relevant data concerning the electrical-equipment industry. Eight members of the thirty-five-firm subsample could not provide estimates, sometimes because they had no experience on which to base such estimates.

TABLE 5.5

*Estimated Annual R and D Expenditure
for Overseas Laboratory of Minimum Economic Scale,
1975: Twenty-seven Firms ($ million)*

INDUSTRY*	SINGLE PRODUCT LINE				SEVERAL PRODUCT LINES			
	RESEARCH	DEVELOPMENT	MINOR PRODUCT CHANGES	RESEARCH AND DEVELOPMENT	RESEARCH	DEVELOPMENT	MINOR PRODUCT CHANGES	RESEARCH AND DEVELOPMENT
Chemicals (n = 7)								
Mean	2.42	3.27	0.72	4.57	2.50	2.46	1.39	3.31
Range	1.0–5.0	0.1–14.0	0.03–2.0	0.46–20.0	1.5–3.5	0.26–6.0	0.24–3.5	0.16–6.7
Petroleum (n = 6)								
Mean	1.64	1.46	0.28	2.30	2.60	2.23	1.18	3.40
Range	0.25–3.0	0.25–3.0	0.15–0.50	0.40–5.0	0.40–5.0	0.40–4.5	0.25–3.0	0.6–7.5
Drugs (n = 2)								
Mean	3.25	0.50	0.12	1.00	6.00	1.30	0.35	†
Range	1.5–5.0	0.5–0.5	0.10–0.13	0.50–1.5	5.0–7.0	0.60–2.0	0.19–0.50	†
Electronics and electrical equipment (n = 5)								
Mean	1.00	1.15	0.40	2.00	2.42	3.95	0.77	6.75
Range	0.8–1.2	0.36–2.0	0.20–0.50	1.25–2.75	1.0–5.0	0.54–8.0	0.51–1.0	1.25–10.0
Glass (n = 2)								
Mean	0.70	0.85	0.42	1.38	0.73	1.53	0.71	1.88
Range	0.1–1.3	0.4–1.3	0.08–0.75	0.50–2.25	0.20–1.25	0.80–2.25	0.16–1.25	1.0–2.75
Total (n = 27)								
Mean	1.58	1.82	0.47	3.00	2.62	2.83	1.23	4.10
Standard deviation	1.34	2.78	0.45	4.10	1.87	2.76	1.30	3.68

Source: see section 6.   *n is the number of firms that provided estimates.   † An estimate was obtained from only one firm.

The results, shown in Table 5.5, indicate that the minimum economic scale tends to be quite substantial in most industries. On the average, for a single product line it was estimated that the expenditures per year for an R and D facility of minimum economic scale would be about $1 million in pharmaceuticals and glass, about $2 million in electrical equipment and petroleum, and about $5 million in chemicals. However, the minimum economic scale seems to vary considerably, depending on the responsibilities of the laboratory. It is less for a laboratory that performs either research or development than for one that performs both, and less for a laboratory that deals with a single product line than for one that deals with several product lines. For a laboratory that is concerned entirely with minor product changes, the average estimated expenditure per year for an R and D facility of minimum economic scale is only about $500,000 per year—and in some industries it is substantially less. In interpreting the results in Table 5.5, the dispersion among the estimates is almost as interesting as the averages. The estimates in each industry vary enormously, reflecting the fact that the minimum economic scale of an R and D laboratory depends on the specific type of work to be done, as well as the fact that opinions differ on this score even among experts.[19]

According to many observers, one major reason why U.S.-based firms have carried out R and D overseas is that costs have tended to be lower there. However, very little information has been published concerning the extent of this cost differential and how it has varied over time. To help fill this gap, we obtained data from the thirty-five-firm subsample concerning the ratio of the cost of R and D inputs in Europe, Japan, and Canada to those in the United States in 1965, 1970, and 1975.[20] The

19. These figures help to explain why, holding other factors constant, smaller firms in equation (5.4) tend to carry on a smaller percentage of their R and D overseas than bigger firms. But they should not be interpreted as saying that smaller firms are squeezed out completely. The estimates in each industry vary enormously. In most industries, at least some of the respondents felt that research and development could be carried out effectively with an annual budget of $500,000, and that minor product changes could be carried out with one of about $100,000. Although these levels of expenditure are hardly trivial, they are within the reach of many firms other than the billion-dollar giants.

Needless to say, these results in no way contradict the finding by many economists that small firms and independent inventors continue to play an important role in the inventive process. Their contribution is frequently in the earlier stages of the inventive process, where the costs are relatively low. Further, according to some observers, costs tend to be lower in smaller organizations, and the figures in Table 5.5 reflect the perceptions of large firms.

20. The relative cost of R and D inputs is the ratio of the annual cost of hiring an R and D scientist or engineer (together with the complementary amount of other inputs) in various overseas locations to do the sort of work carried out there to the annual cost of hiring a comparable R and D scientist or engineer (together with the complementary amount of other inputs) to do the same sort of work in the United States. Each firm was asked to

TABLE 5.6

Mean Ratio of Cost of R and D Inputs in Selected
Overseas Locations to That in the United States,
1965, 1970, and 1975: Thirty-five-Firm Subsample†*

| | LOCATION | | |
YEAR	EUROPE‡	JAPAN	CANADA
1965	0.68	0.56	0.82
1970	0.74	0.60	0.86
1975	0.93	0.90	0.96

*Note that there are many costs of communication and coordination in a multinational network of laboratories.
†Usable data were obtained from nineteen firms. Many of the rest had no overseas R and D experience.
‡There are considerable differences within Europe in the level of R and D costs. According to a number of firms in our sample, costs tended to be relatively low in the United Kingdom and relatively high in West Germany.

results, shown in Table 5.6, indicate that there was a very substantial cost differential in 1965: on the average, the cost of R and D inputs seemed to be about 30 percent lower in Europe, 20 percent lower in Canada, and 40 percent lower in Japan than in the United States. And although there was some increase in R and D costs relative to those in the United States during 1965–70, the cost differential remained quite substantial in 1970.

However, between 1970 and 1975 the situation changed drastically. Owing in part to the depreciation of the dollar relative to other currencies between 1970 and 1975, the cost differential was largely eliminated for many firms. On the average, the cost of R and D inputs was estimated to be about 10 percent lower in Japan and about 5 percent lower in Europe and Canada than in the United States in 1975. Of course, this helps to explain the fact (noted in section 6) that the percentage of R and D carried out overseas was expected to increase less rapidly between 1974 and 1980 than in the period prior to 1974. Since the cost differential between overseas and domestic R and D was smaller, it is quite understandable that firms would expect this percentage to grow less rapidly than in earlier years.[21]

---

estimate this ratio for each year. Many of the estimates were based on studies the firms seem to have carried out in recent years on this topic.

21. If very significant differences exist between the productivity of U.S. and overseas R and D personnel, they may offset the observed differences in the relative costs of inputs. About 80 percent of the firms in our sample regarded the productivity of their R and D personnel in Canada, Europe and Japan to be no lower than those in the United States. Thus, this factor cannot offset the observed difference in the relative cost of R and D inputs in the great majority of firms in our sample.

9. Overseas R and D: Nature of Work and Relation to Domestic R and D

Some observers, as we have seen, are suspicious of overseas R and D because they fear that it may be a channel through which American technology may "leak out" to foreign competitors. The extent to which such a leakage is likely to occur depends in part on the nature of the work being carried out in the overseas laboratories of U.S.-based firms. For example, if such work is focused largely on the modification and adaptation of products and processes for the local market, there is less need to transfer much of the firm's most sophisticated technology overseas than if the work is focused on major product or process developments intended for a worldwide market. Based on information obtained from twenty-three firms in our sample, it appears that these firms' overseas R and D activities tend to focus on development rather than research, on product and process improvements rather than on new products and processes, and on relatively short-term, technically safe work.

Specifically, on the average, the percentage of overseas R and D going for basic research is about six percentage points less than the percentage of domestic R and D going for basic research; the percentage of overseas R and D going for applied research is ten percentage points less than the percentage of domestic R and D going for applied research; while the percentage of overseas R and D going for development is sixteen percentage points greater than the percentage of domestic R and D going for development. Moreover, about three-fourths of these firms' overseas R and D expenditures are aimed at product or process *improvements* and *modifications*, not at entirely new processes or products. This percentage is much higher than for all domestic R and D.

Firms seem to differ considerably in the extent to which they have integrated their overseas R and D with their domestic R and D.[22] Worldwide integration of overseas and domestic R and D exists in almost one-half of the firms (with overseas R and D) in our sample, according to the firms. On the other hand, about one-sixth say that they attempt no such integration, and the rest say that some limited integration is attempted.

Finally, of how much value is overseas R and D to a firm's U.S. operations? Policy makers are interested in this question because it must be considered in any full evaluation of the effects of overseas R and D (and foreign direct investment) on America's technological position vis-à-vis other countries. Unfortunately, practically no evidence exists on this

22. By integration, we mean that the firm's worldwide R and D is viewed as a whole, and laboratories are given worldwide missions, if this seems desirable.

score. To shed a modest amount of light on this question, we obtained estimates from twenty-seven firms in our sample concerning the percentage of their 1975 overseas R and D expenditures with no commercial applicability to their U.S. operations. The results indicate that, on the average, about one-third of these firms' overseas R and D expenditures have no such applicability. Also, we asked each firm to estimate the amount that it would have to spend on R and D in the United States to get results of *equivalent value to its U.S. operations* as a dollar spent overseas. The results, which are only rough, indicate that, on the average, a dollar's worth of overseas R and D seems to result in benefits to these firms' U.S. operations that are equivalent to about fifty cents worth of R and D carried out in the United States.

10. Summary

The cost of designing, constructing, and starting up a manufacturing project abroad based on U.S. technology depends on the time span between project commencement and project completion. If an attempt is made to carry out the project more quickly, it is likely to raise the project's costs. In other words, there is a time-cost tradeoff. In fifteen of the twenty cases in our sample, strategies aimed at shaving the actual time by 1 percent would raise costs by more than 1 percent. Previous work concerning industrial innovations has shown that a 1-percent shaving in project time could be achieved for a cost increase of less than 1 percent in most of the cases examined. Thus, the elasticity of cost with respect to time for the international transfer of manufacturing technology seems greater than for innovation, at least in the samples we studied.[23]

The elasticity of cost with respect to time would be expected to depend on the duration of the preliminary planning stage of the project (relative to other stages), the extent to which the relevant technology has been applied before, the size of the primary transfer agent, the size of the total project costs, and whether or not the foreign market can be satisfactorily supplied by exports. Our econometric results indicate that the effects of all of these factors are statistically significant. If the technology has not been previously commercialized, this elasticity increases by 1.20, on the average. A billion-dollar increase in the sales of the primary transfer agent is associated with a 0.14 increase in this elasticity. A million-dollar increase in the project's total cost is associated with an increase in this elasticity of about the same amount.

23. Obviously, much more data will have to be obtained before we can determine with confidence whether this difference persists. Based on a χ^2 test, this observed difference seems to be significant at about the 0.07 level.

During the 1960s and 1970s, there was a steady and notable increase in the proportion of R and D expenditures carried out overseas by U.S.-based firms. In 1960, the firms in our sample carried out only about 2 percent of their R and D expenditures outside the United States. By 1980, the proportion had increased to over 10 percent.[24] Practically all firms said that the principal reason for doing R and D abroad was to respond to the special design needs of overseas markets. The percentage of a firm's R and D expenditures carried out overseas is directly related to the percentage of its sales derived from abroad and its size. Holding these factors constant, drug firms carry out a relatively high percentage of R and D overseas, due in part to regulations both here and abroad.

According to many observers, one major reason why U.S.-based firms have carried out R and D overseas is that costs have tended to be lower there. In 1965, the cost of R and D inputs seemed to be about 30 percent lower in Europe, 20 percent lower in Canada, and 40 percent lower in Japan than in the United States. But due in part to the depreciation of the dollar, the situation changed drastically during the early 1970s. By 1975, the cost differential had shrunk to only about 5 or 10 percent.

The minimum efficient scale of an overseas R and D laboratory seems to vary considerably, depending on the specific type of work to be done. In general, firms' overseas R and D activities tend to focus on development rather than on research, on product and process adaptation and improvement rather than on new products and processes, and on relatively short-term, technically safe work.[25] Based on very rough estimates, it appears that, on the average, a dollar's worth of overseas R and D may result in benefits to these firms' U.S. operations that are equivalent to about 50 cents' worth of R and D carried out in the United States.

24. In early 1977, the United States Treasury put into effect a new regulation (1.861–8) that, according to some observers, may have tended to increase the amount of R and D done overseas by U.S.-based firms. It required U.S. multinational firms to allocate some of their domestic R and D expenditures against income from foreign sources. Since the estimates in Table 5.4 were made before this new regulation was announced, they do not take this factor into account. Also, they do not take account of recent changes in exchange rates, which may have tended to discourage overseas R and D. However, data published by the National Science Foundation indicate that in 1978 overseas R and D spending by U.S.-based firms amounted to about 10 percent of their company R and D funds. See National Science Foundation, "Greatest Increase in 1978 Industrial R and D Expenditures Provided by 14% Rise in Companies' Own Funds," *Science Resources Studies Highlight*, February 28, 1980. Also see Edwin Mansfield, "Tax Policy and Innovation: Provisions, Proposals, and Needed Research," *National Science Foundation Colloquium on Tax Policy and Investment in Innovation* (Washington, D.C.: National Science Foundation, April 24, 1981). The Economic Recovery Tax Act of 1981 provides for a two-year suspension of regulation 1.861–8. (See Chapter 9 below.)

25. For some relevant discussion, see R. Ronstadt, "R and D Abroad: The Creation and Evolution of Foreign R and D Activities of U.S.-based Multinational Enterprises," Ph.D. diss., Harvard University, 1975.

6 BASIC RESEARCH

AND PRODUCTIVITY INCREASE

IN MANUFACTURING

1. Introduction

As we have seen in previous chapters, international technology transfer has been the focus of considerable controversy in the United States. In part, this is because of the widespread evidence that America's technological lead over its competitors has declined, and the feeling in many quarters that our innovation rate has fallen. In this and the following several chapters, we examine various aspects of the American technological scene that are relevant in this regard and that must be considered in formulating public and private policies to cope with the challenges facing the United States.

More specifically, the present chapter is concerned with the rate of productivity increase in the United States and its relationship to research and development—and particularly basic research. As pointed out in Chapter 1, basic research is defined by the National Science Foundation as "original investigation for the advancement of scientific knowledge ... which do[es] not have immediate commercial objectives."[1] A hotly debated topic among economists, scientists, technologists, and policy makers[2] is: Does basic research, as contrasted with applied research and development, make a significant contribution to an industry's or firm's

1. National Science Foundation, *op. cit.* (cited in Chapter 1, footnote 4).
2. For two influential studies, see Illinois Institute of Technology Research Institute, *op. cit.* (cited in Chapter 1, footnote 9); and Sherwin and Isenson, *First Interim Report on Project Hindsight, op. cit.* (cited in Chapter 1, footnote 9). Of course, the mere fact that industry carries out some basic research does not prove that it makes a significant contribution to innovation and productivity change. Also, see Nelson, "The Simple Economics of Basic Scientific Research" (cited in Chapter 3, footnote 16).

rate of technological innovation and productivity change? Although economic studies indicate that an industry's or firm's R and D expenditures have been directly related to its rate of productivity change, they have been unable to shed light on this question because no attempt has been made to disaggregate R and D. One purpose of this chapter is to begin to fill this void. Specifically, this chapter attempts to determine whether an industry's or firm's rate of productivity change in recent years has been related to the amount of basic research it performed, when other relevant variables (such as its rate of expenditure on applied R and D) are held constant.

A second purpose of this chapter is to present data obtained from 119 firms concerning changes in the composition of their R and D expenditures. In recent years, there has been a widespread feeling that American industry has reduced the proportion of its R and D expenditures going for relatively basic, long-term, and risky projects. However, very few data are available regarding the extent to which such changes in the composition of industrial R and D expenditures have taken place. Our results seem to be the first data on this subject, about which there has been so much discussion.

2. Productivity Growth in the United States

Before describing our methods and results, it is worthwhile to present some basic information concerning productivity growth and R and D expenditures in the United States. Economists and policy makers have long been interested in productivity—the ratio of output to input. The simplest measure of productivity is output per hour of labor, often called labor productivity. Clearly, changes in labor productivity are of fundamental significance, since they are intimately related to, but by no means synonymous with, changes in a country's standard of living. The rate of technological change is one determinant of the rate of growth of labor productivity. Other important determinants of the rate of labor productivity growth are the extent to which capital is substituted for labor, economies of scale, changes in the utilization of productive capacity, and changes in the rate of diffusion of new techniques.

A somewhat more complicated measure of productivity is the total factor productivity index (cited in Chapter 1), which has the advantage that it takes account of both capital and labor inputs. Specifically, this index equals $q \div (zl + vk)$, where q is output (as a percent of output in some base period), l is labor input (as a percent of labor input in some base period), k is capital input (as a percent of capital input in some base period), z is labor's share of the value of output in the base period, and

v is capital's share of the value of the output in the base period. Substituting values of q, l, and k over a given period into this formula, one can easily calculate the value of the index for that period.

Labor productivity in the United States has not increased at a constant rate. The rate of growth of output per man-hour was significantly higher after World War I than before and significantly higher after World War II than before. Specifically, based on John Kendrick's figures, the trend rate of growth of real output per man-hour was almost 2.5 percent for the three decades prior to 1948 (after adjustment for the effect of the Great Depression), as compared with over 3.0 percent from 1948 to 1973. This increase in the rate of growth of labor productivity seems to have been due to a faster increase in real capital per man-hour during 1948–73 than during 1919–48.[3]

After the mid-1960s, however, there has been a notable slowdown in the rate of increase of both output per man-hour and total productivity. Table 6.1 compares the rate of increase of total productivity between 1966 and 1976 with that during 1948–66. In the private domestic business economy, total productivity increased by 1.4 percent per year during 1966–76, as compared with 2.9 percent per year during 1948–66. In most industry divisions (and particularly in mining, construction, and electric and gas utilities), there was a fall in the rate of increase of total productivity.

The data regarding labor productivity show a similar slowdown. According to the Council of Economic Advisers, output per man-hour increased by 1.0 percent during 1973–77, as compared with 2.3 percent during 1965–73. Between 1977 and 1978, it increased by less than one percent. And based on indexes prepared by the Bureau of Labor Statistics, output per hour of labor in the private business sector did not increase at all from 1978 to 1980.

What factors are responsible for this significant slackening of U.S. productivity growth? According to Kutscher, Mark, and Norsworthy of the Bureau of Labor Statistics, one of the major factors has been the increase in the proportion of youths and women in the labor force. Output per man-hour tends to be relatively low among women and among new entrants into the labor force. During the late 1960s, women and new entrants increased as a proportion of the labor force. Based on the bureau's calculations, this change in labor-force composition may have been responsible for 0.2 to 0.3 percentage points of the difference between the average rate of productivity increase in 1947–66 and that

3. Kendrick, "Productivity Trends and Prospects," *op. cit.* (cited in Chapter 1, footnote 21).

TABLE 6.1

Average Annual Rate of Change of Total Productivity,
1948–66 and 1966–76

	1948–66	1966–76
	(*Percent*)	
Private domestic business economy	2.9	1.4
Nonfinancial corporate business	2.7	1.3
Farm	3.5	2.2
Nonfarm-nonmanufacturing	2.5	1.1
Manufacturing	2.5	1.4
Mining	3.3	−1.2
Contract construction	2.5	−1.6
Transportation	2.9	1.6
Communications	4.7	3.3
Public utilities	4.9	−0.3
Trade	2.5	1.4
Finance and insurance	0.4	−0.4
Real estate	3.4	1.8
Services	1.5	1.8

Source: J. Kendrick and E. Grossman, *Productivity in the United States* (Baltimore: Johns Hopkins, 1980).

in 1966–73. George Perry, in his analysis of the causes of the slowdown, agrees that this factor was important.[4]

A second factor that has been cited in this regard is the rate of growth of the capital-labor ratio. During 1948–73, relatively high rates of private investment resulted in a growth of the capital-labor ratio (net nonresidential capital stock divided by aggregate hours worked in the private nonfarm sector) of almost 3.00 percent per year. After 1973, relatively low rates of investment resulted in the growth of the capital-labor ratio by only about 1.75 percent per year. According to the Council of Economic Advisers, this reduction in the rate of growth of the capital-labor ratio may have reduced the rate of productivity increase by up to 0.50 percentage points per year. Christensen, Cummings, and Jorgensen, in their study of this topic, also emphasize the decrease in the rate of growth of the capital-labor ratio.[5]

4. Kutscher, Mark, and Norsworthy, *op. cit.* (cited in Chapter 1, footnote 21); and George Perry, "Labor Structure, Potential Output, and Productivity," in *Brookings Papers on Economic Activity* (Washington, D.C.: The Brookings Institute, 1971).

5. Council of Economic Advisers, *1979 Annual Report* (Washington, D.C.: Government Printing Office, 1979); and Christensen, Cummings, and Jorgensen. *op. cit.* (cited in Chapter 1, footnote 23).

A third factor that has been cited in this regard is increased government regulation. A variety of new types of environmental, health, and safety regulations has been adopted in recent years. Because reduced pollution, enhanced safety, and better health are generally not included in measured output, the use of more of society's resources to meet these regulations is likely to result in a reduction in measured productivity growth. Also, the litigation and uncertainty associated with new regulations may discourage investment and efficiency, and the form of the regulations sometimes may inhibit socially desirable adaptations by firms. According to the Council of Economic Advisers, the direct costs of compliance with environmental health and safety regulations might have reduced the growth of productivity by about 0.4 percentage points per year since 1973.[6]

A fourth factor, cited by Kendrick[7] and others, is the reduction in the rate of increase of intangible capital due to the decrease in the proportion of gross national product devoted to research and development during the late 1960s and early 1970s. In the next section, we shall look closely at the changes over time in the level of R and D expenditures in the United States. For now, it is enough to say that U.S. R and D expenditures decreased, as a percentage of gross national product, from 3.0 percent in 1964 to 2.2 percent in 1978.

A fifth factor, cited by William Nordhaus and others, is the shift of national output toward services and away from goods. However, there is considerable disagreement over whether this shift in the composition of national output is responsible for much of the productivity slowdown. Michael Grossman and Victor Fuchs, among others, are skeptical of this proposition.[8]

Edward Denison has carried out a particularly detailed investigation of the causes of the productivity slowdown.[9] In Table 6.2, we show his estimates of the sources of the growth of national income per person employed (NIPPE). According to these estimates, the advance of knowledge was responsible for 1.4 percentage points of the annual growth rate

6. Council of Economic Advisers, *op. cit.;* and Edward Denison, "Effects of Selected Changes in the Institutional and Human Environment upon Output per Unit of Input," *Survey of Current Business* 58 (1978):21–44.

7. Kendrick, *op. cit.*

8. Wllliam Nordhaus, "The Recent Productivity Slowdown," in *Brookings Papers on Economic Activity* (Washington, D.C.: The Brookings Institution, 1972); and Michael Grossman and Victor Fuchs, "Intersectoral Shifts and Aggregate Productivity Change," in *Annals of Economic and Social Measurement* (New York: National Bureau of Economic Research, 1973).

9. Edward Denison, "The Puzzling Drop in Productivity," *The Brookings Bulletin* (1978); and his *Accounting for Slower Economic Growth* (Washington, D.C.: The Brookings Institution, 1979).

TABLE 6.2

Sources of Growth of National Income per Person Employed,
Nonresidential Business Sector[a]

ITEM	1948–69	1969–73	1973–76	CHANGE FROM 1948–69 TO 1973–76
Growth rate of NIPPE	2.6	1.6	−0.5	−3.1
Effect of irregular factors	−0.1	−0.5	0.1	0.2
Adjusted growth rate	2.7	2.1	−0.6	−3.3
Changes in labor characteristics				
Hours of work	−0.2	−0.3	−0.5	−0.3
Age-sex composition	−0.1	−0.4	−0.3	−0.1
Education	0.5	0.7	0.9	0.4
Changes in capital and land per person employed				
Structures and equipment	0.3	0.2	0.2	−0.1
Inventories	0.1	0.1	0.0	−0.1
Land	0.0	−0.1	0.0	0.0
Improved allocation of resources[b]	0.4	0.1	0.0	−0.4
Changes in legal and human environment[c]	0.0	−0.2	−0.4	−0.4
Economies of scale	0.4	0.4	0.2	−0.2
Advances of knowledge (and not elsewhere classified)	1.4	1.6	−0.7	−2.1

Source: Denison, *op. cit.* See footnote 9.
[a] Detail may not add to totals due to rounding.
[b] Includes only gains due to the reallocation of labor out of farming and out of self-employment in small nonfarm enterprises.
[c] Includes only effects on output per unit of input of costs incurred to protect the physical environment and the safety and health of workers, and of costs of crime and dishonesty.

of NIPPE during 1948–69, and 1.6 percentage points of the annual growth rate of NIPPE during 1969–73. The effects of other factors, such as changes in the characteristics of the labor force, changes in capital and land per person employed, and economies of scale, are also estimated in Table 6.2.

According to Denison, there was a sharp decline in NIPPE during 1974 and 1975. Such declines were without precedent in the period since World War II. Because of them, the 1973–76 rate of growth of NIPPE was negative (− 0.5 percent). When the 1948–69 period is compared with the 1973–76 period, the adjusted rate of growth of NIPPE fell by 3.3 percentage points (from 2.7 percent to − 0.6 percent). Denison's findings indicate that 0.4 percentage points of this decline were due to the use of more resources to meet pollution, safety, and health regulations (and to prevent crime). Another 1.2 percentage points of this decline were attributable to six factors: (1) a steeper drop in working hours; (2)

an accelerated shift in the age-sex composition of employed labor; (3) a slower growth of fixed capital per worker; (4) a slower growth of inventories per worker; (5) reduced gain from resource reallocation (such as the movement to industry of labor overallocated to farming); and (6) reduced gain from economies of scale. Of course, many of these factors have already been cited in earlier paragraphs of this section.

Denison concludes that 2.1 percentage points of the 3.3 point drop in the growth rate of NIPPE remain in the residual called "advances in knowledge and not elsewhere classified." To some extent, this may be due to a slowdown in the rate of technological change and in the rate of innovation. More will be said on this score in Chapter 7. Also, the fact that 1974 and 1975 were the years when big oil price hikes first took effect suggests that their effects may be reflected in the residual. Some observers attribute a substantial proportion of the drop in the rate of productivity increase to the quadrupling of oil prices. Others, like Denison, do not seem to believe that this factor can explain a substantial portion of the decline in the growth rate of NIPPE. The truth is that there is considerable uncertainty regarding the contribution of various factors to the observed productivity slowdown.

3. U.S. Expenditures on R and D, 1953–80

The National Science Foundation has published data for many years concerning the amount spent by industry, government, universities, and others on research and development. Table 6.3 shows total R and D expenditures in the United States from 1953 to 1979. Clearly, R and D expenditures grew very rapidly in the 1950s and early 1960s. In 1953, total R and D spending was about $5 billion, or about 1.4 percent of gross national product. By 1964, total R and D spending was about $19 billion, or about 3.0 percent of gross national product. Industry, impressed by the wartime accomplishments of science and technology, increased its R and D spending greatly. So did the federal government, which poured particularly large sums into defense and space R and D.

The industries that were the leading performers of R and D were aircraft and missiles, electrical equipment, chemicals, motor vehicles, and machinery. These industries accounted for over 80 percent of all R and D performed by industry in 1960, and they continued to do so in 1974. However, it is important to recognize that much of the R and D they performed was (and is) financed by the federal government. During 1964, the government financed about 90 percent of the R and D in the aircraft industry and about 60 percent of the R and D in the electrical equipment industry. (More recent figures on this score will be presented in Table

TABLE 6.3

Expenditures on R and D, United States, 1953–79
(Billions of Dollars)

YEAR	TOTAL R AND D EXPENDITURES (CURRENT DOLLARS)	TOTAL R AND D EXPENDITURES (1972 DOLLARS)	INDUSTRY R AND D EXPENDITURES (1972 DOLLARS)	GOVERNMENT R AND D EXPENDITURES (1972 DOLLARS)
1953	5.1	8.7	3.8	4.7
1955	6.2	10.1	4.1	5.8
1957	9.8	15.1	5.3	9.4
1959	12.4	18.3	6.0	11.9
1961	14.3	20.7	6.9	13.4
1963	17.1	23.9	7.6	15.7
1965	20.1	27.0	8.8	17.5
1967	23.2	29.4	10.3	18.2
1969	25.7	29.6	11.5	17.2
1971	26.7	27.9	11.3	15.6
1973	30.4	28.8	12.2	15.5
1975	35.2	27.7	12.1	14.6
1977	42.9	28.2	12.9	14.3
1979[a]	52.4	32.0	15.1	15.8

Source: National Science Foundation, *National Patterns of R and D Resources, 1953–76* (Washington, D.C.: Government Printing Office, 1976); National Science Foundation, *Science Indicators, 1976* (Washington, D.C.: Government Printing Office, 1977); and National Science Foundation, "Total Federal R and D Growth Slight in 1980 but Varies by Budget Function," *Science Resources Studies Highlights,* June 25, 1979.
[a] Preliminary estimate.

9.4 of Chapter 9.) Table 6.4 shows the intersectoral transfer of funds for R and D in 1980.

In the late 1960s, due partly to the tightening of federal fiscal constraints caused by the Vietnam war, federal expenditures on R and D (in 1972 dollars) decreased. From $18.2 billion in 1967, they fell to $14.5 billion in 1974. Much of this reduction was due to the winding down of the space program and the reduction (in constant dollars) of defense R and D expenditures. During the late 1970s, there once again were increases in constant dollars in federal R and D expenditures. A very large percentage increase occurred in expenditures for energy R and D, which increased (in current dollars) from $1.2 billion in 1975 to $2.9 billion in 1978.

Industry's expenditures on R and D (in 1972 dollars) increased during 1967 to 1978, but at a much slower rate than in 1960–67. In 1960, industry's R and D expenditures (in 1972 dollars) were $6.6 billion; in 1967, they were $10.3 billion; in 1978, they were $14.1 billion. The slower rate

TABLE 6.4

Intersectoral Transfers of Funds for R and D,
United States, 1980 (Billions of Dollars)

| | PERFORMERS | | | | | |
SOURCES OF FUNDS	FEDERAL GOVERNMENT	INDUSTRY	UNIVERSITIES AND COLLEGES	FFRDC'S[a]	OTHER NONPROFIT INSTITUTIONS	TOTAL
Federal government	7.8	14.0	4.1	2.0	1.5	29.4
Industry	—	28.3	0.2	—	0.2	28.7
Universities and colleges	—	—	1.3	—	—	1.3
Other nonprofit institutions	—	—	0.4	—	0.5	1.0
Totals[b] 7.8		42.3	6.0	2.0	2.2	60.4

Source: National Science Foundation, *National Patterns of Science and Technology Resources, 1980,* Washington, D.C.: Government Printing Office, 1980.
[a] Federally funded research and development centers. These are organizations exclusively or substantially financed by the federal government to meet a particular requirement or to provide major facilities for research and training purposes. Those that are administered by industry (such as Oak Ridge National Laboratory or Sandia Laboratory) or nonprofit institutions (such as the RAND Corporation) are included in the respective totals for industry or nonprofit institutions.
[b] Because of rounding errors, items do not always sum to totals.

of increase since 1967 may have reflected a stabilization or decline of the profitability of R and D. For example, Beardsley and Mansfield found, in a study of one of the nation's largest firms, that the private rate of return from its investments in new technology tended to be lower during the late 1960s and early 1970s than during the early 1960s.[10] More will be said on this score in section 6, below.

The nation's total R and D expenditures (including government, industry, and others), when inflation is taken into account, remained essentially constant from 1966 to 1977. The constant dollar figures are very crude, since the National Science Foundation uses the GNP deflator to deflate R and D expenditures. (More will be said on this score in Chapter 10.) But it seems to be generally accepted that no appreciable increase in real R and D expenditures took place during this period. As a percentage of gross national product, R and D expenditures fell from about 3.0 percent in 1964 to about 2.2 percent in 1978. This decline occurred almost continuously from 1964 to 1978, each year's percentage generally being lower than the previous year's percentage.

10. Edwin Mansfield, "Comment," in *New Developments in Productivity Measurement and Analysis,* ed. J. Kendrick and B. Vaccara (Chicago: National Bureau of Economic Research, 1980); and George Beardsley and Edwin Mansfield, "A Note on the Accuracy of Industrial

4. Interindustry Differences in the Rate of Productivity Increase: The Model

Having presented the necessary background material on productivity growth and R and D expenditures in the United States, our next step is to analyze the relationship between an industry's rate of productivity increase and the amount of basic research it performed. The model used here is essentially the same as that employed by Mansfield and Griliches, among others, except that research and development is disaggregated into two parts: (1) basic research and (2) applied research and development. In a particular industry, the production function is assumed to be

$$(6.1) \qquad Q_t = Ae^{\lambda t}R_{tb}^{\alpha_1} R_{ta}^{\alpha_2} L_t^{\nu}K_t^{1-\nu},$$

where Q_t is the industry's value-added in year t, R_{tb} is the industry's stock of basic research capital, R_{ta} is the industry's stock of applied R and D capital, L_t is the industry's labor input, and K_t is the industry's stock of physical capital in year t.[11] Thus, the annual rate of change of total factor productivity[12] is

$$\rho_t = \lambda + \alpha_1 \frac{dR_{tb}/dt}{R_{tb}} + \alpha_2 \frac{dR_{ta}/dt}{R_{ta}}.$$

Forecasts of the Profitability of New Products and Processes," *Journal of Business* 51 (1978):127–35.

11. The concept of a stock of research capital has been used by most investigators in this field. See Mansfield, *Industrial Research and Technological Innovation, op. cit.* (cited in Chapter 1, footnote 1); Zvi Griliches, "Research Expenditures and Growth Accounting," in *Science and Technology in Economic Growth*, ed. B. Williams (London: Macmillan, 1973); Zvi Griliches, "Returns to Research and Development Expenditures in the Private Sector," in *New Developments in Productivity Measurement and Analysis*, ed. J. Kendrick and B. Vaccara (Chicago: National Bureau of Economic Research, 1980); and Nestor Terleckyj, *Effects of R and D on the Productivity Growth of Industries* (Washington, D.C.: National Planning Association, 1974). Our definitions of R_{tb} and R_{ta} are in accord with those used in previous work.

Specifically, the stock of basic research capital in year t equals $\sum_i w_{bi}X_{t-i}$, where X_{t-i} is the expenditure by the industry on basic research in year $t - i$, and w_{bi} is the proportion of this expenditure that is still productive in year t. The stock of applied R and D capital in year t equals $\sum_i w_{ai}A_{t-i}$, where A_{t-i} is the expenditure by the industry on applied R and D in year $t-i$, and w_{ai} is the proportion of this expenditure that is still productive in year t.

12. Because of our specific assumptions concerning the form of the production function, the expression for total factor productivity given here is somewhat different from the one given in section 2. But in both cases, the rate of growth of total factor productivity represents the relative rate of increase (per unit of time) of output when the amounts of capital and labor are held constant.

Since

$$\alpha_1 \frac{dR_{tb}/dt}{R_{tb}} = \frac{\partial Q_t}{\partial R_{tb}} \cdot \frac{dR_{tb}/dt}{Q_t} \text{ and } \alpha_2 \frac{dR_{ta}/dt}{R_{ta}} = \frac{\partial Q_t}{\partial R_{ta}} \cdot \frac{dR_{ta}/dt}{Q_t},$$

(6.2)
$$\rho_t = \lambda + \phi_0 \frac{dR_{tb}/dt}{Q_t} + \phi_1 \frac{dR_{ta}/dt}{Q_t},$$

where

$$\phi_0 = \partial Q_t / \partial R_{tb} \text{ and } \phi_1 = \partial Q_t / \partial R_{ta}.$$

Because the rate of growth of each industry's stock of R and D capital (both basic and applied) cannot be measured directly, we assume, as Griliches and Terleckyj did, that an industry's expenditure on R and D during year t is approximately equal to the change in the industry's stock of R and D capital then. For this assumption to hold, the depreciation of the R and D capital and the lag in the effect of this investment must be small enough to be ignored.[13] Although the assumption of no lags is a useful starting point, we relax this assumption later. (See footnote 17.) According to Kendrick, Terleckyj, and others, an industry's annual rate of change of total factor productivity also depends on the extent to which the industry's workers are unionized.[14] To avoid specification errors, we therefore include u_t, the percent of the industry's workers that are union members, as an additional independent variable in equation (6.2). Thus, the resulting equation is

(6.3)
$$\rho_t = \lambda + \phi_0 \frac{X_t}{Q_t} + \phi_1 \frac{A_t}{Q_t} + \phi_2 u_t,$$

where X_t is the industry's expenditure on basic research in year t, and A_t is the industry's expenditure on applied R and D in year t.

During the period to which our data pertain, there seems to have been only a limited amount of variation in an industry's value of X_t/Q_t, A_t/Q_t, and u_t. Thus, as a first approximation, it seems reasonable to suppose that $X_t/Q_t = B$, $A_t/Q_t = D$, and $u_t = u$. Then if an industry's

13. The rationale given by Griliches and Terleckyj for ignoring depreciation is that, because so little was spent on R and D prior to World War II and because the rate of growth of R and D expenditures in the 1950s was so high, the amount of depreciation that would have been able to build up during the period covered by our data should be small. Relatively little is known about the length of the relevant lags, although some data on this score are presented in Mansfield, *Industrial Research and Technological Innovation, op. cit.*; and Mansfield, "Comment," *op. cit.*

14. See J. Kendrick, *Postwar Productivity Trends in the United States* (New York: National Bureau of Economic Research, 1973); and Terleckyj, *op. cit.* Another variable that Terleckyj found to be statistically significant in some specifications was the percent of the industry's sales to nongovernment buyers. We included this variable in those equations where he found it was significant. For the results, see note a of Table 6.5.

values of λ, ϕ_0, ϕ_1, and ϕ_2 are statistically independent of its values of B, D, and u, it follows that

(6.4) $\rho_i = \bar{\lambda} + \bar{\phi}_0 B_i + \bar{\phi}_1 D_i + \bar{\phi}_2 u_i + z_i,$

where ρ_i is the annual rate of change of total factor productivity in the ith industry during this period; B_i, D_i, and u_i are the values of B, D, and u in the ith industry, $\bar{\lambda}$, $\bar{\phi}_0$, $\bar{\phi}_1$, and $\bar{\phi}_2$ are the average values of λ, ϕ_0, ϕ_1, and ϕ_2 in all relevant industries, and z_i is a random error term.[15]

5. Econometric Results

To what extent can equation (6.4) explain the differences among manufacturing industries in the rate of increase of total factor productivity? To answer this question, we must have data concerning ρ_i, B_i, D_i, and u_i. Kendrick has provided data concerning ρ_i for the twenty two-digit SIC manufacturing industries during 1948–66. Following Terleckyj, who has used a similar sort of model to analyze Kendrick's data, each of the independent variables (other than u_i, which pertains to 1953) is measured as of 1958. To measure B_i, we calculated the ratio of each industry's expenditure on basic research to its value-added in 1958, based on data published by the National Science Foundation and the Commerce Department.[16] Terleckyj has provided data concerning D_i and u_i. Using these data, we regressed ρ_i on B_i, D_i, and u_i, with the results shown in Table 6.5. All three independent variables turn out to be highly significant, and over 60 percent of the variation among industries in the rates of productivity increase are explained by the equation.[17]

15. With regard to A_t/Q_t and X_t/Q_t, the earliest comparable data are for 1953. Comparing an industry's value of A_t/Q_t in 1953 with that in 1966, we find that there is a high correlation ($r = .91$) between them, and that the intertemporal variation in A_t/Q_t is much smaller than the interindustry variation in it. Similar results were found for X_t/Q_t.

16. The data regarding B_i are based on the basic research expenditures reported in the 1961 edition of the National Science Foundation, *Research and Development in Industry* (Washington, D.C.: Government Printing Office, 1961). Note that both government-financed and company-financed basic research are included. In two cases, the National Science Foundation provided data regarding two of our industries combined. We allocated the basic research expenditures between the two industries in proportion to their applied R and D expenditures (as reported in Terleckyj). This procedure is very unlikely to result in any serious error, since these industries do very little basic research. If an industry did no applied R and D in 1958, according to Terleckyj, and the National Science Foundation provided no indication that it carried out any basic research, we assumed that its basic research expenditures were zero in 1958.

17. The effects of R and D often occur with a lag, and the lag for basic research is generally thought to be much longer than for applied R and D. Because so little is known about the length of these lags, previous work along this line has sometimes (but not always) ignored them; but once we separate basic research from applied R and D, it is particularly

To see how robust these results are, we experimented with some other specifications of the regression equation. First, we divided the total amount of applied R and D into two parts: the amount that is privately financed, and the amount that is financed by government. This means that two independent variables—D_{pi} (the amount of privately financed applied R and D divided by value-added) and D_{Gi} (the amount of government-financed applied R and D divided by value-added)—must be substituted for D_i. As shown in Table 6.5, this substitution has relatively

---

hazardous to continue this procedure, since the lags in the effects of basic research may be quite long. If an expenditure on basic research is included in the stock of basic research capital only after a lag of θ_1 years, and if an expenditure on applied R and D is included in the stock of applied R and D capital only after θ_2 years, equation (6.3) must be modified as follows:

$$\rho_t = \lambda + \phi_0 \frac{X_{t-\theta_1}}{Q_t} + \phi_1 \frac{A_{t-\theta_2}}{Q_t} + \phi_2 u_t.$$

If r_1 is the average annual rate of growth of the industry's value-added between year $t - \theta_1$ and year t, and if r_2 is the average annual rate of growth of the industry's value-added between year $t - \theta_2$ and year t, it follows that

$$\rho_t = \lambda + \phi_3 \frac{X_{t-\theta_1}}{Q_{t-\theta_1}} + \phi_4 \frac{A_{t-\theta_2}}{Q_{t-\theta_2}} + \phi_2 u_t,$$

where $\phi_3 = \phi_0 \div (1 + r_1)^{\theta_1}$ and $\phi_4 = \phi_1 \div (1 + r_2)^{\theta_2}$. If the independent variables in this equation remain relatively constant in the relevant period, and if an industry's values of λ, ϕ_2, ϕ_3, and ϕ_4 are statistically independent of the values of these independent variables,

$$\rho_t = \bar{\lambda} + \bar{\phi}_3 B_i'' + \bar{\phi}_4 D_i'' + \bar{\phi}_2 u_i + z_i,$$

where B_i'' is the value of $X_{t-\theta_1} / Q_{t-\theta_1}$ and D_i'' is the value of $A_{t-\theta_2} / Q_{t-\theta_2}$ in the ith industry during the relevant period.

Although we do not know θ_1 and θ_2 with much accuracy, it is very likely that there is a high correlation between B_i'' and B_i, as well as between D_i'' and D_i. For as far back as reasonably accurate data exist, industries with relatively high ratios of basic research to value-added in one year have tended to have relatively high ratios in other years, and industries with relatively high ratios of applied R and D to value-added in one year have tended to have relatively high ratios in other years. The coefficient of correlation between an industry's ratio of applied R and D expenditures to value-added in 1953 and its ratio in 1975 is 0.93. The coefficient of correlation between an industry's ratio of basic research expenditures to value-added in 1953 and its ratio in 1975 is 0.85. Even if the value of θ_1 or θ_2 varies from industry to industry, there is likely to be a high correlation between B_i'' and B_i and between D'' and D_i.

Thus, B_i and D_i can be regarded as proxies for B_i'' and D_i''. Viewed in this way, our regression equations, since they are based on these proxies rather than on B_i'' and D_i'', cannot be used to estimate the rates of return from basic research or from applied R and D, unless one can estimate the unknown values of θ_1 and θ_2, as well as the relationship between these proxies and the variables for which they stand. However, the results are useful in testing whether ϕ_0 equals zero, since, if this is the case, the regression coefficient of B_i in our regressions should be zero. Thus, despite the presence of these lags, our results are useful in testing whether an industry's or firm's rate of productivity change has been affected by the amount of basic research it performed, when its rate of expenditure on applied R and D is held constant.

TABLE 6.5

Results of Regressions to Explain Rates of Productivity
Increase, Twenty Manufacturing Industries, 1948–66 [a]

INDEPENDENT VARIABLES [b]

CONSTANT	B_i	D_i	U_i	D_{pi}	D_{gi}	I_i	I_{pi}	Y_i	$\bar{R}^2$
4.80	1.49 (3.34)	0.068 (2.75)	−0.056 (3.90)	—	—	—	—	—	0.63
4.74	1.33 (1.51)	—	−0.055 (3.54)	0.105 (0.59)	0.055 (0.80)	—	—	—	0.63
4.37	1.66 (4.55)	−0.084 (1.59)	−0.055 (4.73)	—	—	0.54 (3.10)	—	—	0.71
4.28	1.50 (4.94)	−0.18 (0.71)	−0.054 (5.57)	—	—	—	0.74 (4.41)	—	0.80
4.46	0.285 (0.81)	0.084 (5.17)	−0.061 (6.37)	—	—	—	—	0.065 (4.80)	0.86
4.83	0.682 (1.27)	—	−0.064 (6.32)	−0.023 (0.21)	0.122 (2.88)	—	—	0.068 (4.90)	0.87
4.46	0.572 (1.67)	0.005 (0.14)	−0.059 (7.02)	—	—	0.27 (2.07)	—	0.054 (4.16)	0.86
4.42	0.665 (1.85)	0.035 (1.29)	−0.058 (6.81)	—	—	—	0.37 (2.11)	0.045 (2.90)	0.86

[a] Besides these regressions, we calculated all of Terleckyj's equations in his Tables 4–9 with B_i or both B_i and Y_i added (and with the nonsignificant non–R and D variables in his equations omitted). In practically all cases, the regression coefficient of B_i is significant when it alone is added. When both B_i and Y_i are added, the regression coefficient of Y_i is always highly significant, but the regression coefficient of B_i is significant in only one-third of the equations. The estimated coefficients of the independent variables are generally similar to those shown here.
[b] All of the independent variables, as well as the dependent variable, are expressed as percentages. The number in parentheses below each regression coefficient is its *t*-value.

little effect on the estimate of ϕ_0, although it reduces its statistical significance.

Second, we added to equation (6.4) a fourth independent variable, I_i, which measures the amount of R and D that is embodied in an industry's purchased inputs (divided by its value added). Terleckyj has provided estimates of this variable for these twenty industries. As shown in Table 6.5, when this variable is added, it is highly significant, but its inclusion has little effect on the size or statistical significance of the estimate of $\bar{\phi}_0$. Similarly, if this variable is defined to include only privately financed R and D, the results are essentially the same, as shown in Table 6.5 (where this variable is denoted by I_{pi}).

These findings seem to indicate a strong relationship between the amount of basic research carried out by an industry and the industry's rate of productivity increase during 1948–66. However, one wonders whether, since the distinction between basic research and applied research is often nebulous, this relationship may not reflect the fact that industries that carry out relatively large amounts of long-term R and D tend to have relatively high rates of productivity increase. In other words, basic research may be a proxy for long-term R and D. To shed some light on this question, we added another variable, Y_i (the percent of the industry's firms that expected their R and D expenditures in 1958 to pay off in no less than six years), to the four regressions discussed above.[18] As shown in Table 6.5, the results suggest that much of the apparent effect of basic research may really be due to long-term R and D. When this additional variable is introduced, its effect is highly significant in every case, whereas the effect of B_i becomes smaller and often statistically nonsignificant.

In addition, to extend the time period to which the data pertain, we used unpublished data by Kendrick to calculate ρ_i for these twenty manufacturing industries during 1948–76. When these values of ρ_i are used in Table 6.5, the magnitude of the regression coefficient of B_i—and the number of specifications where it is statistically significant—is about the same as in Table 6.5. However, the regression equations do not fit nearly as well, probably due in part to the increased importance during 1966–76 of regulatory considerations and a variety of other factors not included in the model. Also, the regression coefficients of Y_i are no longer statistically significant. This is not surprising because there have been considerable changes since 1958 (the year to which Y_i pertains) in the extent to which various industries have focused their R and D on long-term work. Much more will be said on this score in section 7.[19]

18. Data concerning Y_i were obtained from McGraw-Hill, *Business' Plans for New Plants and Equipment*, 1958–61, p. 8. The figure for primary metals is the average of that given for "iron and steel" and for "nonferrous metals." The figure for transportation equipment is the average of that given for "autos" and that given for "transportation equipment." The figure for fabricated metal products and instruments is that given for "other metalworking." Since three of the industries carried out no applied R and D expenditures in 1958, according to Terkeckyj, no data regarding Y_i exist for these industries.

19. John Kendrick kindly provided me with estimates of total factor productivity in each industry during 1948–76. These estimates were to appear in Kendrick and Grossman, *op. cit.* We calculated the annual percentage rate of growth of total factor productivity between 1948 and the year since 1966 when total factor productivity reached a maximum. (If total factor productivity is less than in a previous year, the industry is probably not operating on its production function, and the inclusion of such an observation may result in misleading results.) The year when it reaches a maximum is almost always 1976 or 1973. When the resulting values of ρ_i are inserted in the regressions in Table 6.5, the regression coefficient of B_i is statistically significant at the .05 level in about the same proportion of cases as in Table 6.5.

6. Interfirm Differences in the Rate of Productivity Increase

Besides investigating the relationship between basic research and productivity increase at the industry level, we also investigate this relationship at the firm level. Once again, we assume that equation (6.1) holds, but now Q_t is the firm's value-added in year t, R_{tb} is the firm's stock of basic research capital, R_{ta} is the firm's stock of applied R and D capital, and so forth. Making the same assumptions as in section 4, it follows that

$$(6.5) \qquad \rho_i' = \bar{\lambda}' + \bar{\phi}_0' B_i' + \bar{\phi}_1' D_i' + z_i',$$

where ρ_i' is the annual rate of change of total factor productivity in the ith firm; B_i' and D_i' are the ratio of basic research expenditure to value-added and the ratio of applied R and D expenditure to value-added in the ith firm; and $\bar{\lambda}'$, $\bar{\phi}_0'$, $\bar{\phi}_1'$, and z_i' are analogous (at the firm level) to $\bar{\lambda}$, $\bar{\phi}_0$, $\bar{\phi}_1$, and z_i. Equation (6.5) is analogous to equation (6.4) except that u_i is excluded because it has the same value for all firms in the industry.[20]

To see how well this model fits, and to estimate $\bar{\phi}_0'$ and $\bar{\phi}_1'$, we obtained data concerning value-added and employment during 1960–76 and concerning investment in plant and equipment during 1920–76 for ten major petroleum firms and for six major chemical firms.[21] The stock of physical capital was estimated annually for each firm during 1960–76. To obtain such an estimate for a particular year, the investment data for each firm for previous years were deflated, depreciated (using the Jack Faucett–BLS depreciation rates for plant and equipment in chemicals and petroleum refining), and summed. Using these data concerning each

20. Of course, we could have included a variable measuring interfirm differences in the extent of unionization, but this did not seem worthwhile because the interfirm differences in this regard seemed small.

21. The data concerning value-added and employment were obtained from *Moody's Industrials* and from the annual reports of the firms. The figures regarding value-added were deflated by the Bureau of Economic Analysis's implicit deflator for the industry. Most of the data concerning investment in plant and equipment come from annual reports and *Moody's*. However, for the earliest years, it was often necessary to estimate a firm's investment in plant and equipment during certain years by assuming it was proportional to the change in the firm's total assets. Although these estimates for the early years are crude, the results should be influenced very little by whatever errors they contain, since the investment figures for the early years have relatively little weight in the capital stock figures. This is because most of the plant and equipment acquired in the early years was depreciated by 1960. In the case of the petroleum firms, the analysis was carried only up to 1971, not 1976, because of the gyrations in oil prices and the unsettled conditions in the industry in recent years.

firm's stock of physical capital each year, together with the data concerning each firm's (deflated) value-added and employment each year, we estimated the annual rate of total factor productivity change for each firm during 1960–76.[22] In addition to these estimates of ρ_i', we need estimates of B_i' and D_i' for each of these firms. As a by-product of previous research, we had these data for all of these firms for 1964 or for 1965.[23]

To estimate the parameters of equation (6.5), we regressed each firm's value of ρ_i' on its values of B_i' and D_i', the result being

(6.6) $\rho_i' = 2.15 + 1.78B_i' + 0.10D_i',$ $(\bar{R}^2 = 0.58)$
                   (7.51) (2.08) (0.86)

where the t-statistics are shown in parentheses below the regression coefficients, and where both the dependent and independent variables are expressed as percentages. To determine whether there were interindustry differences between the chemical and petroleum firms, we introduced as an additional independent variable a dummy variable that equals 1 if the firm is in the chemical industry and 0 if it is in the petroleum industry. This dummy variable was not statistically significant, so it was omitted from equation (6.6).[24]

The results are surprisingly similar to those obtained at the industry level. There is a statistically significant tendency for a firm's rate of pro-

22. The investment figures were deflated by price deflators provided by the Bureau of Labor Statistics. Then the Faucett-BLS depreciation rates were applied, and the results were summed to obtain the firm's capital stock each year. Then, for each firm, we computed $g_t = ln\,Q_t - b\,ln\,L_t - (1 - b)\,lnK_t$, where Q_t is the firm's deflated value-added in year t, L_t is its employment in year t, K_t is its capital stock in year t, and b is its average proportion of value-added going for wages during 1960–76. Finally, we regressed g_t on t, the resulting regression coefficient being the estimate of ρ_i'. Years when the firm's value-added was less than in some previous year were omitted because the firm was apparently operating well below its capacity. Clearly, in an analysis of this sort, the data should apply to periods when the firm is operating reasonably close to capacity. The data described in this and the previous footnote were obtained, and the calculations were made, by Nancy Fox as part of her work toward a doctoral dissertation at the University of Pennsylvania.

23. Some of these data were described in Mansfield, Rapoport, Schnee, Wagner, and Hamburger, *Research and Innovation in the Modern Corporation op. cit.* (cited in Chapter 1, footnote 6). The rest come from an unpublished survey carried out by Henry Armour and David Teece. To check whether the ratio of applied R and D expenditures to value-added remained relatively constant during the relevant period, we compared the values of this ratio in 1976 and (about) 1964 for the six chemical firms. In most firms, the 1976 ratio was within 25 percent of the 1964 ratio. The interyear differences tend to be much smaller than the interfirm differences. There was a high correlation between a firm's 1976 ratio and its 1964 ratio. With regard to the ratio of basic research expenditures to value-added, the results are much the same.

24. The same result was obtained in Mansfield, *op. cit.* (cited in footnote 11) on the basis of earlier data.

ductivity increase during 1960–76 to be directly related to how much basic research it carried out in 1964, when the amount spent on applied research and development is held constant. Moreover, the regression coefficient of B_i' in equation (6.6) is not too different from the regression coefficient of B_i in equation (6.4).

Previous studies have used such regression coefficients as estimates of the private rate of return from research and development.[25] For well-known reasons, some of which are discussed in footnote 17, estimates of this sort should be treated with considerable caution. Nonetheless, it is interesting to see how an estimate of this sort based on these data compares with previous estimates, since so little is known about the changes that have occurred in recent years in the profitability of R and D. (Practically all previous work of this kind pertains to the early 1960s.) To obtain such an estimate, we regress ρ_i' on $(B_i' + D_i')$, the result being

$$(6.7) \qquad \rho_i' = 2.20 + 0.275\,(B_i' + D_i'), \quad (\bar{R}^2 = .49)$$
$$(7.06) \quad (4.08)$$

which implies that the rate of return from all R and D (basic and applied) was about 27 percent in 1960–76 in these chemical and petroleum firms. This estimate is lower than corresponding estimates (using the same technique and the same industries) for the early 1960s, a result that is consistent with the reduced rate of growth of R and D expenditures by industry during the early 1970s and with opinions expressed by many R and D managers.[26]

7. Recent Changes in the Composition of Industrial R and D Expenditures

In recent years, there has been a widespread feeling that industry has been devoting a smaller share of its R and D expenditures to basic research, long-term projects, and risky and ambitious projects. Unfortunately, however, little or no data have been available on this score. To make some progress toward filling this gap, we obtained information from 119 firms concerning the changes that have occurred in this regard between 1967 and 1977, and the changes they expected between 1977

25. For example, Griliches, *op. cit.* (1980).

26. Basic research and applied R and D are reaggregated in equation (6.7) so that the results will be comparable with earlier studies. In 1960, the estimated rate of return in these industries averaged about 30 percent or more, according to Mansfield, *op. cit.* (cited in footnote 11). In the early 1960s, it was about 40 percent, according to Griliches, *op. cit.* (1980). Of course, all of these estimates neglect lags, and should be treated with considerable caution.

and 1980. The firms included in our sample, all of which spent over $10 million on R and D in 1976, accounted for about one-half of all industrial R and D expenditures in the United States in 1976. Although this sample has its shortcomings, it is certainly among the most comprehensive surveys of this type ever carried out.[27]

The results, presented in Table 6.6, indicate that the proportion of R and D expenditures devoted to basic research declined between 1967 and 1977 in practically every industry. In the aerospace, metals, electrical equipment, office equipment and computer, chemical, drug, and rubber industries, this proportion dropped substantially. In the sample as a whole, the proportion fell about one-fourth, from 5.6 percent in 1967 to 4.1 percent in 1977. According to the firms' forecasts, there was no evidence that this drop would continue during 1977–80, but neither was there any evidence that the proportion would rise very much. For the sample as a whole, the proportion was expected to be about 4.3 percent in 1980.[28]

In four-fifths of the industries, there was also a decline between 1967 and 1977 in the proportion of R and D expenditures devoted to relatively risky projects (specifically, ones with less than a fifty-fifty estimated chance of success). In some industries, like metals, chemicals, aircraft, drugs, and rubber, this reduction has been rather large. Between 1977

27. Data were obtained (through correspondence or interviews) from practically all of the major chemical firms, from about three-quarters of the major aerospace, automotive, petroleum, metals, and office equipment and computer firms, from about two-thirds of the major rubber, soap, and machinery firms, from about one-half of the major drug, food, and instruments firms, and from about one-quarter of the major electronics and electrical equipment firms. By a major firm, we mean one that spent over $10 million on R and D in 1976, according to the survey in *Business Week,* June 27, 1977. The data were obtained in practically all cases from the firm's vice-president of research and development. One obvious limitation of the sample is that it focuses entirely on large firms. For an interesting recent study which deals with some of the topics discussed here, see H. Nason, J. Steger, and G. Manners, *Support of Basic Research by Industry* (Washington, D.C.: National Science Foundation, 1978).

28. According to NSF's figures, 4.7 percent of company-financed R and D expenditures in manufacturing went for basic research in 1975 (the most recent year for which NSF had published data). This is reasonably close to our 1977 figure of 4.1 percent. In 1968, NSF put this percentage at 6.9 percent, whereas our 1967 figure is 5.6 percent. For nine industries in 1977 and seven industries in 1967, we could compare our percentage with NSF's percentage. We found that our percentages are positively correlated with NSF's, but that they do not always agree closely. This is not surprising, since our definition of each of these industries differs somewhat from NSF's, and NSF's coverage is more comprehensive than ours. Nonetheless, there is close agreement with regard to the change in this percentage in all manufacturing. NSF's data indicate a decrease of 32 percent in this percentage between 1968 and 1975, and our data indicate a decrease of 27 percent between 1967 and 1977. For the NSF data, see National Science Foundation, *Research and Development in Industry,* which was published for both 1975 and 1968.

TABLE 6.6

Percentage of Company-financed R and D Expenditure Going for Basic, Long-Term, and Relatively Risky Projects, 1967, 1977, and 1980[a]

INDUSTRY	BASIC RESEARCH			PROJECTS LASTING FIVE OR MORE YEARS			PROJECTS AIMED AT ENTIRELY NEW PRODUCTS AND PROCESSES			PROJECTS WITH LESS THAN FIFTY-FIFTY ESTIMATED CHANCE OF SUCCESS		
	1967	1977	1980	1967	1977	1980	1967	1977	1980	1967	1977	1980
Metals	6.2	2.4	2.4	26	22	22	28	18	19	18	11	12
Chemicals	7.3	5.9	5.8	43	39	39	37	33	31	37	30	27
Aerospace	8.5	1.7	1.7	31	24	24	35	26	28	35	27	29
Automobiles	0.3	0.2	0.3	18	20	23	15	16	a	47	45	45
Petroleum	9.1	8.1	9.7	26	27	29	21	25	26	24	23	24
Drugs	20.7	16.4	15.3	63	66	63	76	68	65	46	40	38
Food	2.6	4.1	4.3	16	16	22	18	22	25	19	18	20
Instruments	3.7	3.3	3.3	40	40	41	24	22	22	11	9	10
Soap and cosmetics	3.4	3.9	3.6	23	30	37	38	48	55	43	55	55
Machinery	4.2	3.6	3.4	34	31	32	23	18	20	8	5	5
Electronics and electrical equipment	6.4	4.6	4.8	10	14	14	37	45	47	3	5	5
Office equipment and computers	3.5	1.8	a	55	55	a	60	60	a	9	9	a
Stone, clay, and glass	8.0	7.1	8.1	35	43	38	37	28	27	28	24	21
Rubber	2.7	1.7	1.7	9	5	4	32	28	28	37	27	22
Miscellaneous	5.1	2.5	3.5	30	27	27	21	22	27	18	13	17
Totals	5.6	4.1	4.3	34	34	a	36	34	34	28	25	27

[a]The 1980 figures are the firms' forecasts. Some 1980 figures are not entirely comparable with earlier ones because some firms did not provide 1980 forecasts. In cases where the lack of comparability is serious, no 1980 figure is given.

and 1980, there was expected to be some overall increase in the proportion devoted to relatively risky projects, with the result that the average
proportion for the sample as a whole was expected to get closer to its
1967 level.

Despite the decrease in the proportion of R and D expenditures
devoted to basic research, the proportion devoted to relatively long-term
projects (specifically, ones lasting five or more years) did not decline
appreciably between 1967 and 1977 in the sample as a whole. In some
industries, like aircraft, chemicals, metals, and rubber, there was a substantial decline; but in other industries, like drugs, there was an increase
in this proportion. However, according to a number of firms, this is misleading because it is due largely to regulatory and related factors. If we
omit the industries (like drugs) where firms stated that regulatory changes
resulted in a significant lengthening of projects, there was about a 4 percent decline in the average proportion of R and D expenditures devoted
to relatively long-term projects. In most industries, the firms expected
this proportion to increase somewhat between 1977 and 1980.[29]

The proportion of R and D expenditures aimed at entirely new
products and processes (rather than improvements and modifications of
existing products and processes) declined somewhat between 1967 and
1977.[30] Again, there are marked differences among industries in the
amount and direction of change. Between 1977 and 1980, little change
was expected to occur in this proportion in most industries.

A multiple regression analysis provides no evidence of any relationship between a firm's size, profitability, or percentage of R and D expenditures devoted to sales, on the one hand, and the change in the
proportion of the firm's R and D expenditures devoted to basic research
between 1967 and 1977 (or expected to occur between 1977 and 1980),
on the other. Moreover, there is no relationship between these three
variables and the change in the proportion of the firm's R and D expenditures devoted to relatively risky projects, or to relatively long-term
projects, or to projects aimed at entirely new products and processes.[31]

29. Other data regarding the proportion of firms' R and D expenditures that goes
for relatively long-term projects are presented in Mansfield et al., op. cit., and Mansfield,
op. cit. (cited in footnote 10). These data are quite consistent with the results in Table 6.6.

30. The distinction between an entirely new product or process and an improved or
modified product or process is often arbitrary (although it is often used). So long as each
respondent uses the same definition for one year as for another year, the results should
help to portray the changes that have occurred over time in a given industry. But a comparison in this regard of one industry with another may be misleading since the definitions
used in one industry may be quite different from those used in another industry.

31. We ran four regressions, the dependent variables being (a) the change between
1967 and 1977 in the proportion of a firm's R and D expenditures devoted to basic research,
(b) the change between 1967 and 1977 in the proportion devoted to projects where there

Further, there is only a very weak relationship, if any, between the change in the proportion of a firm's R and D expenditures devoted to basic research and the change in the proportion devoted to relatively risky projects, relatively long-term projects, or projects aimed at entirely new products or processes.[32]

Besides obtaining data from these 119 firms, we also asked those that have cut back on the proportion of their R and D expenditures going for basic research and relatively risky and long-term projects why such a change has occurred. The reason most frequently given was the increase in government regulations, which has reduced the profitability of more fundamental and longer-term projects. This reason was advanced particularly often by the chemical and drug firms.[33] Another reason advanced by some of the respondents was that breakthroughs are more difficult to achieve than in the past, because the field has been more thoroughly worked over. Still another frequently cited factor was the inflation in recent years.[34]

8. Summary

From 1948 to 1966, labor productivity increased at a relatively high rate, relative to previous years. But in 1966–73, it increased more slowly, and in 1973–80, it increased at an even slower pace. This productivity slowdown is one of the most important problems facing the American economy. To some extent, it seems to have been due to changes in the age and sex composition of the labor force, reductions in the rate of growth of the capital-labor ratio, and increases in government regula-

---

was less than a fifty–fifty estimated chance of success, (c) the change between 1967 and 1977 in the proportion devoted to projects lasting five or more years, and (d) the change between 1967 and 1977 in the proportion aimed at entirely new products or processes. Each of these dependent variables was regressed on the firm's sales in 1976, its ratio of R and D expenditures to sales in 1976, its ratio of profits to net worth in 1976, and industry dummy variables. Other than some of the dummy variables, none of the independent variables in any of these multiple regressions was statistically significant.

32. We regressed three dependent variables—the change between 1967 and 1977 in the proportion of a firm's R and D expenditures devoted to projects lasting five or more years, the change in the proportion aimed at entirely new products or processes, and the change in the proportion devoted to projects where there was less than a fifty–fifty estimated chance of success—on the change between 1967 and 1977 in the proportion devoted to basic research. Although the sample size exceeded one hundred, in only one case was the regression statistically significant, and in this case, r^2 was only .06.

33. About 80 percent of the relevant chemical firms and practically all of the relevant drug firms cited this reason.

34. See Edwin Mansfield, "R and D, Productivity, and Inflation," (cited in Chapter 1, footnote 46).

tion. Another factor is the reduction in the rate of increase of intangible capital due to the lack of growth of real R and D expenditures between 1965 and 1977. Since 1974, the big hikes in oil prices may also have been a factor. Unfortunately, although economists have devoted appreciable study to the subject, considerable uncertainty exists regarding the contribution of various factors to the observed productivity slowdown.

During the 1950s and early 1960s, R and D expenditures grew very rapidly in the United States. During the late 1960s and early 1970s, federal expenditures on R and D decreased in real terms, due to the winding down of the space program and the reduction (in constant dollars) of defense R and D. Industry's R and D expenditures increased during 1967 to 1978, but at a much slower rate than in 1960–67. The nation's total R and D expenditures (including government, industry, and others), when inflation is taken (roughly) into account, remained essentially constant from 1966 to 1977.

Using a simple model of the sort employed previously by Mansfield, Griliches, and Terleckyj, among others, we analyzed the effects of both basic research and applied R and D on an industry's rate of productivity growth. When its expenditures on applied R and D are held constant, there is a statistically significant and direct relationship between the amount of basic research carried out by an industry and its rate of increase of total factor productivity. Whether the relevant distinction is between basic and applied research is by no means clear; there is some evidence that basic research may be acting as a proxy for long-term R and D.[35] Holding constant the amount spent on both applied R and D and basic research, an industry's rate of productivity increase during 1948–66 seems to be directly and significantly related to the extent to which its R and D was long-term. This appears to be the first systematic evidence that the composition, as well as the size, of an industry's R and D expenditures affects its rate of productivity increase.

When the same sort of model is applied to data for individual firms in the chemical and petroleum industries, the results once again indicate a significant relationship between basic research and productivity growth. If the customary sort of estimate of the private rate of return from R and D is calculated, based on data for 1960–76, it turns out to be 27 percent, which is lower than corresponding estimates (using the same techniques and the same industries) for the early 1960s. This reduction in the profitability of R and D may have been partly responsible for the

35. It is possible too that industries and firms with high rates of productivity growth tend to spend relatively large amounts on basic research, but that their high rates of productivity growth are not due to these expenditures. Much more work is required before we can be entirely confident of the way in which this observed relationship should be interpreted.

lower rate of increase of industry's R and D expenditures since the late 1960s than before.

The composition of many industries' R and D expenditures has changed considerably since the late 1960s. Practically all industries have cut the proportion of their R and D expenditures going for basic research. Most industries have cut the proportion going for relatively ambitious and risky projects. When all industries are combined, the changes seem less major than when the focus is put on particular industries, like chemicals, drugs, aerospace, metals, and rubber. In each of these industries, noteworthy changes have occurred. Some of the factors responsible for these changes are increased regulation, higher rates of inflation, and bigger apparent difficulties in obtaining technical breakthroughs.

7 PATENTS, IMITATION COSTS, AND THE RATE OF INNOVATION

1. Introduction

In practically any discussion of public policy toward technological change, the patent system is sure to be cited. It is an old and basic part of America's public policy in this area. Nonetheless, the fact that it is old does not mean that it cannot be improved or that its effects are well understood. In this chapter, we describe some of the proposed and recent changes in the patent system. Also, we provide some of the first empirical data concerning the effects of patents on imitation costs and the rate of innovation.

As Kenneth Arrow and others have pointed out, if firms can imitate an innovation at a cost that is substantially below the cost to the innovator of developing the innovation, there may be little or no incentive for the innovator to carry out the innovation.[1] In their discussions of the innovation process, economists frequently have called attention to the major role played by the costs of imitation, but there has been little or no attempt to measure these costs, to test various hypotheses concerning the factors influencing them, or to estimate their effects. In this chapter, we report some findings of what seems to be the first study of this topic.

2. The Patent System

One of the major instruments of national policy regarding technology is the patent system. The United States patent laws grant the inventor exclusive control over the use of his invention for seventeen years in

1. K. Arrow, "Economic Welfare and the Allocation of Resources for Invention," in National Bureau of Economic Research, *The Rate and Direction of Inventive Activity* (New York: National Bureau of Economic Research, 1962).

exchange for his making the invention public knowledge. Not all new knowledge is patentable. A patentable invention "is not a revelation of something which existed and was unknown, but the creation of something which did not exist before."[2] "There can be no patent upon an abstract philosophical principle."[3] A patentable invention must have as its subject matter a physical result or a physical means of attaining some result, not a purely human means of attaining it. Moreover, it must contain a certain minimum degree of novelty. " 'Improvement' and 'invention' are not convertible terms. ... [W]here the most favorable construction that can be given ... is that the article constitutes an improvement over prior inventions, but it embodies no new principle or mode of operation not utilized before by other inventors, there is no invention."[4]

Since Congress passed the original patent act in 1790, the arguments used to justify the existence of the patent laws have not changed very much. First, these laws are regarded as an important incentive to induce the inventor to put in the work required to produce an invention. Particularly in the case of the individual inventor, it is claimed that patent protection is a strong incentive. Second, patents are regarded as a necessary incentive to induce firms to carry out the further work and make the necessary investment in pilot plants and other items that are required to bring the invention to commercial use. If an invention became public property when made, why should a firm incur the costs and risks involved in experimenting with a new process or product? Another firm could

2. Pyrene Mfg. Co. v. Boyce, C.C.A.N.J., 292 F. 480.

3. Boyd v. Cherry, 50 F. 279, 282.

4. William Schwarzwaelder and Co. v. City of Detroit, 77 F. 886, 891. What proportion of patented inventions are used commercially, and how great are the private returns from them? The Patent, Copyright, and Trademark Foundation of George Washington University selected a 2-percent random sample of patents granted in 1938, 1948, and 1952 and determined (by interviews or correspondence with the inventor or assignee) the utilization status of the inventions sampled. According to the results, 51 percent of the patents assigned at date of issue to large companies were used commercially, 71 percent of those assigned at date of issue to small companies were used commercially, and 49 percent of the patents unassigned at date of issue were used commercially. For mechanical inventions, 57 percent were used commercially; whereas for electrical and chemical inventions, 44 percent were used commercially.

Not all of the patents in commercial use are profitable to their owners. The Patent, Copyright, and Trademark Foundation obtained information on profitability from assignees on 127 of 292 inventions reported in past or current use. Two-thirds of those used in the past (for which there were data) were profitable; one-third showed losses. Nine-tenths of those currently in use (for which there were data) were profitable; one-tenth showed losses. Dollar estimates of profits and losses were provided by assignees for 93 patents in current or past use. For the results, see B. Sanders, "Patterns of Commercial Exploitation of Patented Inventions by Large and Small Companies," *Patent, Copyright, and Trademark Journal* (1964).

watch, take no risks, and duplicate the process or product if it were successful. Third, it is argued that, because of the patent laws, inventions are disclosed earlier than otherwise, the consequence being that other inventions are facilitated by the earlier dissemination of the information. The resulting situation is often contrasted with the intense secrecy with regard to processes which characterized the medieval guilds and which undoubtedly retarded technological progress and economic growth.[5]

Not all economists agree that the patent system is beneficial. A patent represents a monopoly right, although as we shall see in section 8, it may represent a very weak one. Critics of the patent system stress the social costs arising from the monopoly. They point out that, after a new process or product has been discovered, it may cost little or nothing for other persons who could make use of this knowledge to acquire it. (However, as stressed in Chapter 4, the cost of technology transfer frequently is substantial.) The patent gives the inventor the right to charge a price for the use of the information, the result being that the knowledge is used less widely than is socially optimal.[6] Critics also point out that patents have been used to create monopoly positions that were sustained by other means after the original patents had expired; they cite as examples the aluminum, shoe machinery, and plate glass industries.[7] In addition, the cross licensing of patents often has been used by firms as a vehicle for joint monopolistic exploitation of their market.

Critics also question the extent of the social gains arising from the system. They point out that the patent system was designed for the individual inventor, but that over the years most research and development has become institutionalized. They assert that patents are not really important as incentives to the large corporation, since it cannot afford to fall behind in the technological race, regardless of whether or not it receives a patent. They also assert that, because of long lead times, most

5. See F. Machlup, *An Economic Review of the Patent System*, Study 15 of the Senate Subcommittee on Patents, Trademarks, and Copyrights, 1958.

6. Since the use of knowledge by one individual does not reduce the ability of another individual to use it, it would be socially desirable in a static sense for existing knowledge to be available for use wherever it is of social value, if it costs nothing to transfer it. (But as stressed throughout this book, transfer costs frequently are substantial.) However, if there were no financial reward for the knowledge producer, there would be less incentive to produce new knowledge. To promote optimal use while preserving this incentive, it has been suggested that society would be better off if the profits obtained by the patent holder could be awarded to him as a lump sum and if there were unrestricted use of the information. Unfortunately, this proposal suffers from important practical difficulties. See E. Mansfield, "Economics, Public Policy and the Patent System," *Journal of the Patent Office Society* (1965); and E. Mansfield, "National Science Policy," *American Economic Review* 56 (1966):476–88, as well as section 3 below.

7. See A. Kahn, "The Role of the Patents," in J. Miller, *Competition, Cartels, and Their Regulation* (Amsterdam: North Holland, 1962).

of the innovative profits from some types of innovations can be captured before imitators have a chance to enter the market. Moreover, they claim that firms keep secret what inventions they can, and patent those that they cannot.[8]

Surveys of business firms provide mixed answers about the importance of patents in encouraging R and D and innovation. Some firms feel that patents are extremely important and that their research and innovative activities could not be sustained without them; others feel that patents make little difference in their search for, and introduction of, new products and processes. The drug and chemical industries make extensive use of patents; the automobile, paper, and rubber industries do not. Patents are generally regarded as more important for independent inventors and small firms than for large firms, which are better able to carry out their inventive and innovative activities without the protection of the patent system. They seem to be more important for major product inventions than for process inventions (which can often be kept secret for a considerable period of time) or for minor product inventions.[9]

Do the benefits derived from the patent system outweigh its costs? Like many broad issues of public policy, the facts are too incomplete and too contaminated by value judgments to permit a clear-cut, quantitative estimate of the effects of the patent system. Nonetheless, with or without such an estimate, it is impossible to avoid the relevant policy issue, and when confronted with it, there are few leading economists, if any, who favor abolition of the patent system. Even those who publish their agnosticism with respect to the system's effects admit that it would be irresponsible, on the basis of our present knowledge, to recommend abolishing it. Nonetheless, many economists seem to be interested in altering some of its features—which is hardly surprising in view of the fact that the patent system has continued for over a century without really major change.

3. Proposed and Recent Changes in the Patent System

Over the years, there have been many proposals for reforming the U.S. patent system. Some have proposed that the length of the patent be varied in accord with the importance of the invention or the cost of mak-

---

8. See C. Edwards, *Maintaining Competition: Requisites of a Government Policy* (New York: McGraw-Hill, 1949); A. Plant, "The Economic Theory Concerning Patents for Inventions," *Economica* (1934); and Machlup, *op. cit.*

9. See Kahn, *op. cit.*; F. Scherer *et al., Patents and the Corporation* (Boston: Harvard Business School, 1959); and the National Academy of Sciences—National Research Council, *The Role of Patents in Research* (Washington, D.C., 1962).

ing it. This proposal seems reasonable, since it is highly improbable that a single inflexible system can apply optimally both to major technological breakthroughs and to more routine, incremental changes. However, there are administrative difficulties arising from the problem of deciding the relative importance of inventions.[10]

Turning to more radical proposals, a system of general compulsory licensing has sometimes been suggested, which would permit everyone to obtain licenses under any patent. Under this system, patentees could no longer hope for attractive monopoly profits, but only for royalties from their licenses and cost advantages over their royalty-paying competitors. This proposal has been resisted almost everywhere because of the difficulties of determining "reasonable royalties" and because of a fear that inventive and innovative activity would be unduly discouraged.[11] Another proposal, put forth by Michael Polanyi, would "supplement licenses of right by government rewards to patentees on a level ample enough to give general satisfaction to inventors."[12] Every inventor would have the right to claim a public reward, and all other persons could use the invention freely. Because of the difficulty in setting individual awards, this proposal has received only a limited amount of attention.

It seems unlikely that fundamental changes of this sort will be made. If changes come about, they are likely to be less sweeping. For example, drug firms, among others, have argued that a patent owner should be allowed to receive an extension of a patent's life equal to the length of government regulatory delays.

In addition, there is a question as to whether or not we in the United States are correct in allowing employee-inventors to assign away all their rights in patents relating to the business of the employer or resulting from work done for the employer, in consideration of employment. In West Germany, for example, the employed inventor receives under certain circumstances an extra compensation for his invention. Very little is

10. See Machlup, *op. cit.;* and P. Federico, *Renewal Fees and Other Patent Fees in Foreign Countries,* Study 17 of the Senate Subcommittee on Patents, Trademarks, and Copyrights, 1958.

11. An important facet of the controversy over compulsory licensing is the charge of suppression of patents which has been persistently repeated and angrily rejected. See Floyd Vaughn, *The United States Patent System* (Norman: University of Oklahoma Press, 1956); Edwards, *op. cit.;* and M. Polanyi, "Patent Reform," *Review of Economic Studies* (1944). Also, see C. Corry, *Compulsory Licensing of Patents—a Legislative History,* Study 12 of the Senate Subcommittee on Patents, Trademarks, and Copyrights, 1958.

12. M. Polanyi, *ibid.* Proposals for systems of prizes and bonuses to inventors, rather than patents, are very old. For example, Alexander Hamilton suggested that the federal government award prizes for important inventions. However, the problem of selecting inventors and inventions has limited the attention accorded this sort of scheme.

known about the effects of such schemes on the performance of the employee-inventor or on the employer's incentives to carry out research and development. In the United States, there is little indication that existing practices will be altered in the foreseeable future.[13]

In December 1980, Congress passed an act amending the patent laws. According to this act, any person can seek a reexamination by the Patent and Trademark Office of a patent. According to its proponents, reexamination will allow efficient resolution of questions concerning the validity of issued patents without recourse to long and costly infringement litigation. Also, the fee structure for patents was revised, and the fees are to be paid in a number of installments over the life of the patent. If the invention has no commercial value, the inventor can let the patent lapse, and pay only those installments that already have come due.

In addition, this act tries to move toward a uniform policy toward inventions made with federal support. During the past thirty years, government patent policy has been a topic of acrimonious debate. Some agencies, like the Department of Defense, have allowed the title to the patents to remain with the contractor; other agencies have retained title to the patents. The 1980 act stipulates that small businesses and nonprofit organizations normally are to have the right to elect to retain worldwide ownership of their inventions, although under some circumstances other arrangements may be made. The government may require small businesses and nonprofit organizations to license such inventions if reasonable efforts are not being made to achieve practical application.[14]

4. Patent Statistics and the Rate of Invention

The number of patents is sometimes used as a crude index of the rate of invention in a given field during a particular period of time. Used in this way, patent statistics have important disadvantages. For one thing, the average importance of the patents granted at one time and place may differ from those granted at another time and place. For another, the proportion of the total inventions that are patented may vary significantly. Nonetheless, it is worthwhile examining the behavior of postwar patent statistics at least briefly.

In Table 7.1, we show the changes over time in the number of U.S. patents granted. The number of patents granted to U.S. residents rose

13. For example, see F. Neumeyer, *The Law of Employed Inventors in Europe*, Study 30 of the Senate Subcommittee on Patents, Trademarks, and Copyrights, 1963.

14. Some of these revisions were related to suggestions of the Domestic Policy Review on Industrial Innovation, described in Chapter 9.

TABLE 7.1

U.S. Patents Granted to U.S. Residents,
by Year of Grant, 1960–78

YEAR	NUMBER OF PATENTS (THOUSANDS)	YEAR	NUMBER OF PATENTS (THOUSANDS)
1960	39.5	1969	50.4
1961	40.2	1970	47.1
1962	45.6	1971	56.0
1963	37.2	1972	51.5
1964	38.4	1973	51.5
1965	50.3	1974	50.6
1966	54.6	1975	46.7
1967	51.3	1976	44.3
1968	45.8	1978	41.2

Source: National Science Foundation, *Science Indicators, 1980* (Washington, D.C.: Government Printing Office, 1981).

during the 1960s, reached a peak in 1971, and was about 20 percent lower in 1976 than in 1971. When patents are broken down by product field, the results are much the same. For example, the number of patents on electrical equipment, instruments, and communication equipment all reach a peak in 1971. In some product fields like machinery and fabricated metals, the peak is reached in the late 1960s. In chemicals, it is reached in 1972.

The number of patents granted is a measure of inventive activity in a previous period, since roughly two years are taken by the Patent Office to process and examine a patent application. To correct for this, Table 7.2 shows the number of patents by year of application. There is much less year-to-year fluctuation in the number of patents when application dates rather than grant dates are used. And, as would be expected, the peak patenting rate now occurs in 1969, rather than 1971. When the data are broken down by product field, the results are surprisingly uniform. In machinery, fabricated metals, and electrical equipment, the peak is in 1966; in communication equipment and chemicals, it is in 1969; and in instruments, it is in 1971. In practically all of the fifty-two product fields for which data are available, the patent rate declined during the 1970s. The only exceptions are in drugs, agricultural chemicals, and motorcycles, bicycles, and parts.

Thus, the patent statistics seem to be quite consistent with the view (cited in the previous chapter) that the rate of invention has been slowing down in the United States. However, the crudeness of patent statistics as a measure of the rate of invention should be emphasized. Although

TABLE 7.2

U.S. Patents Due to U.S. Inventors,
by Year of Application, 1965–73 [a]
(Thousands of Patents)

	1965	1966	1967	1968	1969	1970	1971	1972	1973
All patents	42.2	45.0	44.1	45.3	46.3	45.6	45.3	41.9	41.6
Food	0.5	0.5	0.5	0.5	0.5	0.6	0.6	0.4	0.4
Textiles	0.4	0.4	0.5	0.5	0.5	0.5	0.5	0.4	0.3
Chemicals	5.8	6.2	6.1	6.1	6.3	6.2	6.1	5.4	4.9
Drugs	0.6	0.7	0.6	0.6	0.7	0.7	0.7	0.7	0.7
Oil and gas	0.6	0.7	0.7	0.8	0.8	0.8	0.7	0.6	0.6
Rubber	2.3	2.4	2.4	2.3	2.5	2.3	2.3	2.2	1.8
Stone, clay, and glass	1.0	1.0	1.0	1.0	1.1	1.1	1.1	1.0	0.9
Primary metals	0.5	0.5	0.5	0.5	0.5	0.6	0.5	0.5	0.5
Fabricated metals products	5.8	6.4	6.2	6.2	6.2	6.1	6.0	5.6	5.1
Machinery	13.3	14.3	13.8	14.0	14.2	13.9	13.9	12.7	11.7
Electrical equipment	4.8	5.3	5.1	5.2	5.1	5.1	4.8	4.5	4.3
Communications and electronics	5.2	5.4	5.1	5.2	5.7	5.5	5.3	5.0	4.7
Transportation equipment	2.2	2.6	2.5	2.7	2.6	2.6	2.7	2.5	2.5
Aircraft	0.5	0.7	0.7	0.7	0.8	0.7	0.7	0.7	0.7
Instruments	4.5	4.6	4.7	4.8	5.0	5.1	5.2	4.8	4.7

Source: National Science Foundation, op. cit.
[a] Most of these patents were assigned to U.S. corporations. Note too that the decline
in patent applications continues beyond 1973. See Table 7.1 and its source.

these statistics should not be ignored, neither should a disproportionate amount of emphasis be put on them. Also, it should be noted that the decline in the patent rate is not unique to the United States. There has been such a decline in West Germany during 1973–76; in France during 1971–76; in the United Kingdom during 1971–76; and in Canada during 1971–76. (In each country, we include only patents granted to the country's nationals, not all patents.)[15]

5. Imitation Costs and Times

Patents generally increase imitation costs. That is, they increase the amount that an imitator must spend to imitate a new process or product that is patented. Imitation costs play a very important role in the innovation and diffusion processes. For example, as pointed out in section 1, the costs of imitating new products have an important effect on the incentives for innovation in a market economy. Unfortunately, however, despite the fact that many economists have cited the need for empirical studies of imitation costs and imitation times, no such studies have been carried out.

To help fill this gap, we obtained data from firms in the chemical, drug, electronics, and machinery industries concerning the cost and time of imitating (legally) forty-eight product innovations. By imitation cost we mean all costs of developing and introducing the imitative product, including applied research, product specification, pilot plant or prototype construction, investment in plant and equipment, and manufacturing and marketing start-up.[16] (If there was a patent on the innovation, the cost of inventing around it is included.) By imitation time we mean the length of time elapsing from the beginning of the imitator's applied research (if there was any) on the imitative product to the date of its commercial introduction.[17]

The firms were chosen more or less at random from among the major firms in these four industries in the Northeast, and the new products were chosen more or less at random from among those introduced recently by these firms.[18] In thirty-four of the cases, the new product

15. *Science Indicators, 1978* (Washington, D.C.: National Science Foundation, 1979).

16. For definitions of these terms, see Mansfield, Rapoport, Schnee, Wagner, and Hamburger, *Research and Innovation in the Modern Corporation, op. cit.* (Cited in Chapter 1, footnote 6.)

17. For many products, the estimate of imitation cost (and of imitation time) was an average of two estimates provided by the firms, each reflecting a somewhat different set of assumptions.

18. We say that the firms and innovations were chosen "more or less" at random because a table of random numbers really was not used. In some cases, the sample was a

had already been imitated, so the data are based on actual experience. In the remaining fourteen cases, no imitator had appeared as yet, but the innovating firm provided us with detailed estimates that were regarded as being reliable.[19] Also, in all forty-eight cases, data were obtained from the innovating firm concerning the costs of the innovation, as well as the time it took to bring the innovation to market (from the beginning of applied research to the date of its commercial introduction).

The innovations included in the sample vary widely in importance, but practically all are major new products that are central to the innovators' activities, not peripheral to their main business. For thirty of the forty-eight products, the innovation cost exceeded $1 million; for twelve products, it exceeded $5 million. About 70 percent of the innovations were patented. In only one case did the innovator license the imitating firm. In all other cases, the imitator received no help from the innovator. In general, these firms did not appear to be very interested in licensing these innovations, at least in the relevant time period, often because they did not feel it was in their interest to encourage potential rivals.

On the average, the ratio of the imitation cost to the innovation cost was about 0.65, and the ratio of the imitation time to the innovation time was about 0.70. As shown in Tables 7.3 and 7.4, there is considerable variation about these averages. In about half of the cases, the ratio of imitation cost to innovation cost was either less than 0.40 or more than 0.90. In about half of the cases, the ratio of imitation time to innovation time was either less than 0.40 or more than 1.00. Products with a rela-

---

systematic sample from a list; in other cases, there were so few innovations in the time frame we selected that the choice was automatic. Both successful and unsuccessful new products were included in the sample. About half of the innovations turned out to be successful (in the sense that they were relatively profitable to the innovator). A high percentage of the firms we approached agreed to cooperate so there was little problem of nonresponse. A considerable number of in-depth interviews were held with major officials of each firm to insure that the data were as accurate as possible. Seventeen of the innovations occurred in the chemical industry (including petroleum refining); eighteen occurred in the drug industry; ten occurred in the electronics industry; and three occurred in the machinery industry. Five of the new products were first introduced before 1960; fifteen were first introduced during the 1960s; and twenty-eight were first introduced during 1970–76.

19. In each such case, the innovating firm provided an estimate of how much money and time it would have cost the most likely (and most efficient) imitator to have imitated the product. In some cases, it was assumed that the time-cost combination that would have been chosen by the imitator was midway between the least-cost combination and the least-time combination. According to the firms, these estimates are unlikely to be very wide of the mark. If we had dropped these cases from the sample, we would have risked the introduction of a serious bias. Thus, this seemed to be the best feasible procedure. This procedure also was used in other cases and seemed accurate enough for present purposes.

TABLE 7.3

Imitation Cost (Divided by Innovation Cost)
of Forty-eight New Products, by Industry and Cost of Innovation

IMITATION COST (DIVIDED BY INNOVATION COST)	INNOVATIONS COSTING MORE THAN $1 MILLION			INNOVATIONS COSTING LESS THAN $1 MILLION		
	CHEMICALS	DRUGS	ELECTRONICS AND MACHINERY	CHEMICALS	DRUGS	ELECTRONICS AND MACHINERY
A. Number of New Products						
Less than 0.20	1	1	1	1	0	0
0.20 and under 0.40	0	3	0	0	0	2
0.40 and under 0.60	1	1	2	5	0	0
0.60 and under 0.80	2	5	0	2	0	4
0.80 and under 1.00	2	3	1	2	1	1
1.00 and over	2	2	1	1	0	1
Totals	8	15	5	11	1	8
B. New Products Weighted by Innovation Cost (Percent)[a]						
Less than 0.20	3	3	17	15	0	0
0.20 and under 0.40	0	11	0	0	0	34
0.40 and under 0.60	b	1	53	46	0	0
0.60 and under 0.80	44	54	0	13	0	36
0.80 and under 1.00	15	21	9	22	100	18
1.00 and over	38	9	22	4	0	11
Totals[c]	100	100	100	100	100	100

[a] The weighted number of new products is expressed as a percentage of the column total.
[b] Less than 0.5.
[c] Because of rounding errors, items may not sum to column total.

tively high (low) ratio of imitation cost to innovation cost tended to have a relatively high (low) ratio of imitation time to innovation time.[20]

It may come as a surprise that imitation cost was no smaller than innovation cost in about one-seventh of the cases. This was not due to any superiority of the imitative product over the innovation. Instead, in a substantial percentage of these cases, it was due to the innovator's having a technological edge over its rivals in the relevant field. Often this edge was due to superior "know-how"—that is, better and more extensive technical information based on highly specialized experience with the development and production of related products and processes. Such know-how is not divulged in patents and is relatively inaccessible (at least for a period of time) to potential imitators.

20. The correlation coefficient between a product's ratio of imitation cost to innovation cost and its ratio of imitation time to innovation time is about .8.

TABLE 7.4

Imitation Time (Divided by Innovation Time)
of Forty-eight New Products, by Industry and Cost of Innovation

IMITATION TIME (DIVIDED BY INNOVATION TIME)	INNOVATIONS COSTING MORE THAN $1 MILLION			INNOVATIONS COSTING LESS THAN $1 MILLION		
	CHEMICALS	DRUGS	ELECTRONICS AND MACHINERY	CHEMICALS	DRUGS	ELECTRONICS AND MACHINERY
	A. Number of New Products					
Less than 0.30	1	2	1	2	0	0
0.30 and under 0.50	1	5	1	2	0	2
0.50 and under 0.70	1	3	1	4	1	2
0.70 and under 0.90	3	0	0	1	0	1
0.90 and under 1.10	1	3	1	1	0	1
1.10 and over	1	2	1	1	0	2
Totals	8	15	5	11	1	8
	B. New Products Weighted by Innovation Cost (Percent)[a]					
Less than 0.30	3	10	35	16	0	0
0.30 and under 0.50	2	28	17	36	0	16
0.50 and under 0.70	35	24	19	27	100	23
0.70 and under 0.90	22	0	0	16	0	14
0.90 and under 1.10	18	16	22	4	0	18
1.10 and over	19	23	9	2	0	30
Totals[b]	100	100	100	100	100	100

[a] The weighted number of new products is expressed as a percentage of the column total.
[b] Because of rounding errors, items may not sum to column total.

Based on these results, it appears that innovators routinely introduce new products despite the fact that other firms can imitate these products at about two-thirds (often less) of the cost and time expended by the innovator.[21] In some cases, this is because, although other firms could imitate these products in this way, there are other barriers to entry (for example, lack of a well-known brand name) that discourage potential imitators. But to a greater extent (at least in this sample), it seems to be due to a feeling on the part of the innovators that, even if imitators do begin to appear in a relatively few years, the innovation still will be profitable.

21. The available data pertain only to innovations that were introduced, not to those where the incentives were insufficient for their introduction. Although they cannot measure the effects of imitation cost on how many innovations were introduced, they provide valuable evidence concerning the characteristics of those that were introduced.

6. The Imitator's Time-Cost Tradeoff Function

Clearly, the time it takes a firm to imitate a new product can generally be reduced by spending more money. Each product's imitator was confronted by a time-cost tradeoff function, which is the relationship between the amount spent by the imitator and the length of time it would take to imitate this new product.[22] The time-cost combination chosen by the imitator is one point on this tradeoff function. In most cases, we obtained an estimate of the minimum time that the imitator could have taken to carry out the project, and the corresponding cost, as well as an estimate of the minimum cost and the corresponding time. Together with other information, this enabled us to determine the approximate position of the imitator's time-cost tradeoff function for most products. If imitation cost is measured as a percentage of innovation cost and if imitation time is measured as a percentage of innovation time, all of these time-cost tradeoff functions can be plotted on the same graph. When this is done, we find that a product's ratio of imitation cost to innovation cost is a good indicator of how high and how far to the right the product's time-cost tradeoff function is located.

Using the time-cost tradeoff function for each product, we computed the arc elasticity of cost with respect to time. The median value of this elasticity is about 0.7, which means that a 1-percent reduction in time results in about a 0.7 increase in cost, on the average.[23] These elasticities pertain to the imitation time that is midway between minimum time and the time corresponding to minimum cost. Since the time-cost tradeoff function is convex, this elasticity increases as imitation time falls (and approaches its minimum value), and decreases as imitation time rises (and approaches the time corresponding to minimum cost). For most products, the minimum-time imitation cost is less than 50 percent greater than the minimum imitation cost, but in some cases the difference is 100 percent or more. Ordinarily, the minimum-cost imitation

22. For further discussion of time-cost tradeoff functions, see Chapter 5. In addition, see F. M. Scherer, "Research and Development Resource Allocation under Rivalry," *Quarterly Journal of Economics* 81 (1967):359–94; F. M. Scherer, *Industrial Market Structure and Economic Performance,* 2d ed. (Chicago: Rand McNally, 1980); Mansfield *et al., op. cit.;* and W. Baldwin and G. Childs, "The Fast Second and Rivalry in Research and Development," *Southern Economic Journal* 36 (1969):18–24.

23. Estimates of the elasticity of cost with respect to time could be obtained for thirty-nine products. In five cases, it was less than 0.25; in ten cases, it was 0.25 and under 0.50; in twelve cases, it was 0.50 and under 1.00; in ten cases, it was 1.00 and under 2.00; and in two cases, it was 2.00 and over.

time is less than 75 percent greater than the minimum imitation time, but in over one-fourth of the cases the difference is 100 percent or more (Table 7.5).

TABLE 7.5

*Ratio of Minimum-Time Imitation Cost to Minimum
Imitation Cost, and Ratio of Minimum-Cost
Imitation Time to Minimum Imitation Time,
Percentage Distribution of New Products in the Sample*

VALUE OF RATIO	MINIMUM-TIME IMITATION COST ÷ MINIMUM IMITATION COST	MINIMUM-COST IMITATION TIME ÷ MINIMUM IMITATION TIME
	(Percent of New Products)	
1.00 and under 1.25	36	18
1.25 and under 1.50	33	24
1.50 and under 1.75	6	24
1.75 and under 2.00	6	3
2.00 and under 2.25	12	15
2.25 and under 2.50	6	6
2.50 and over	0	9
Totals	100[a]	100[a]

[a] Because of rounding errors, the percentages do not sum to the column total.

7. Determinants of Imitation Costs

The following hypotheses help to explain the substantial variation among products in the ratio of imitation cost to innovation cost.[24] First, we would expect the ith product's ratio of imitation cost to innovation cost (denoted by C_i) to be inversely related to the proportion of the ith product's innovation cost that goes for research (rather than product

24. These hypotheses also help to explain the variation among products in the ratio of imitation time to innovation time, since (as pointed out in footnote 20) a product's ratio of imitation time to innovation time is positively correlated with its ratio of imitation cost to innovation cost.

For some purposes, one might want to explain the variation among products in the ratio of imitation cost to innovation cost when the ratio of imitation time to innovation time is held constant (at its mean value). Based on our estimates of the time-cost tradeoff functions, we were able to make rough estimates of most products' ratio of imitation cost to innovation cost under these circumstances. When these estimates are used (in place of C_i) in equation (7.2), the results are relatively unchanged. Thus, the hypotheses discussed in this section can explain the variation among products in both the unadjusted ratio of imitation cost to innovation cost and the ratio of imitation cost to innovation cost when the ratio of imitation time to innovation time is held constant.

At first glance, one might suppose that whether or not an innovation was patented

specification, pilot plant or prototype, plant and equipment, or manu-facturing or marketing start-up). An imitator frequently can spend much less time and money on research than the innovator because the prod-uct's existence and characteristics provide the imitator with a great deal of information that the innovator had to obtain through its own research. On the other hand, an imitator often has to go through many of the same steps as the innovator with respect to pilot plant or prototype con-struction, investment in plant and equipment, and manufacturing and marketing start-up.

Second, we would expect C_i to be relatively large if the ith product is a new drug where the Food and Drug Administration requires that an imitative product must be tested in much the same way as the innovative product with which it competes. As Kitch and others have pointed out, this requirement increases the money and time that an imitator must spend. (How much it increases them is a matter of some debate; results on this score are given below.) Third, we would expect C_i to be relatively small if the ith product consists of a new use for an existing material and if some firm other than the innovator has patents on this material. In such cases, the patent holder often can imitate the innovation relatively cheaply and quickly.[25]

To test these hypotheses, we assume that

$$(7.1) \qquad C_i = \alpha_0 + \alpha_1 R_i + \alpha_2 D_i + \alpha_3 G_i + z_i,$$

where R_i is the percentage of the innovation cost that went for applied research in the case of the ith product, D_i is a dummy variable that equals 1 if the ith innovation was an ethical drug (subject to regulation of the

---

should be included as an independent variable in equation (7.1). But there is a problem here. Some innovations are patented because the ratio of imitation cost (without the pat-ent) to innovation cost is low, while some innovations are not patented because the ratio of imitation cost (without the patent) to innovation cost is high. Thus, the ratio of imita-tion cost to imitation cost may not be very different among patented innovations than among unpatented innovations. In fact, for this sample, when a dummy variable (which equaled 1 if the innovation was patented and zero otherwise) was added to equation (7.1), it was not statistically significant. However, this does not mean that patent protection does not influence the ratio of imitation cost to innovation cost. In section 8, we look in detail at this effect of patents.

25. In one case in our sample, an agricultural chemical firm found a new use for a particular chemical. This firm's supplier held a patent on the broad group of chemicals of which this chemical was part, but was unaware of this new use. When the innovator mar-keted the chemical for this new use, the supplier could (and did) imitate the innovation quickly and at a low cost.

Another factor we would like to include in equation (7.1) is the extent of the imitator's technological capabilities and know-how, relative to those of the innovator, but we could find no way to measure this factor adequately.

sort described above) and 0 otherwise, G_i is a dummy variable that equals 1 if the ith product was a new use for an existing material on which a firm other than the innovator holds patents and 0 otherwise, and z_i is a random error term. Data were collected regarding R_i, D_i, and G_i for twenty-eight of the products in our sample,[26] and we computed least-squares estimates of the α's. The results are

$$(7.2) \quad C_i = 0.838 - 0.00684R_i + 0.310D_i - 0.536G_i. \quad (\bar{R}^2 = .49)$$
$$\quad\quad (11.8) \quad\quad (3.80) \quad\quad (2.31) \quad\quad (2.41)$$

The t-ratios are given in parentheses.

Each of the regression coefficients in equation (7.2) has the expected sign, is statistically significant, and is large enough so that the effect of each independent variable is substantial. If R_i increases by twenty percentage points, C_i decreases by about 0.14. If the ith product is a new drug, C_i is about 0.31 larger than would otherwise be the case. If the ith product is a new use for an existing material on which a firm other than the innovator holds the patents, C_i is about .54 smaller than would otherwise be the case. These three independent variables can explain about half of the variation in C_i.

8. Patents and Imitation Costs

As pointed out in section 5, still another factor that affects the ratio of imitation cost to innovation cost is whether or not the innovator has patents on the new product. Contrary to popular opinion, patent protection does not make entry impossible, or even unlikely. Within four years of their introduction, 60 percent of the patented successful innovations in our sample were imitated. Nonetheless, patent protection generally increases imitation costs. To obtain information concerning the size of this increase, the firms in our sample were asked to estimate how much the value of C_i for each patented product increased because it was patented. The median estimated increase in C_i was 11 percent. They also were asked to estimate how much the value of C_i for each unpatented product would have increased if it had been patented. The median esti-

26. The sample was composed of two subsamples, each of which was collected quite independently from the other. In the subsample for which Schwartz was particularly responsible, we obtained the data needed to test these hypotheses. In the other subsample, for which Wagner was particularly responsible, we did not obtain these data, but focused attention on other questions instead. Equation (7.4) is also based only on the former subsample. One product was not included in equation (7.2) because the imitator received a license from the innovator, but if this product is included, the results do not change in any significant way. Interaction terms were also included in equation (7.2), but were not significant.

mated increase in C_i was only about 6 percent. (Indeed, for two of these products, patent protection would have reduced the money and time required for imitation because in these cases the innovator could keep secret the essential information underlying the product, whereas if it patented it, some of the information would have been disclosed.) The fact that a patent resulted in a larger increase in the imitation costs of the patented products than of the unpatented products was, of course, a major reason why some products were patented and others were not.[27]

In the ethical drug industry, patents had a bigger impact on imitation costs than in the other industries, which helps to account for survey results[28] indicating that patents are regarded as more important in ethical drugs than elsewhere. The median estimated increase in C_i due to patent protection was about 30 percent in ethical drugs, in contrast to about 10 percent in chemicals and about 7 percent in electronics and machinery. Without patent protection, it frequently would have been relatively cheap (and quick) for an imitator to determine the composition of a new drug and to begin producing it. However, for many of these electronics and machinery innovations, it would have been quite difficult for imitators to determine from the new product how it is produced, and patents would not add a great deal to imitation cost (or time).[29] These results are in accord with Taylor and Silberston's conclusion that the lack of patent protection would reduce the rate of expenditure on innovative activity to a greater extent in drugs than in other industries.

27. Since some of the products in the sample were inherently unpatentable, they had to be omitted, so the results reported in this paragraph are based on forty-three products. Using the Wilcoxon test, the difference between the median increase in C_i for patented and unpatented products is significant at the .05 probability level. In the entire sample of forty-eight products, the percentage that was patented was 81 (drugs), 63 (chemicals), and 69 (electronics and machinery).

28. For a description of these surveys and their results, see C. Taylor and Z. Silberston, *The Economic Impact of the Patent System* (Cambridge: Cambridge University Press, 1973); and F. M. Scherer, *The Economic Effects of Compulsory Patent Licensing* (New York: New York University Monograph 1977–2 in Finance and Economics, 1977).

29. Taylor and Silberston, *ibid.*, point out that a high proportion of patents in electronics are thought to be of doubtful validity, and that, even if an electronics patent is valid, it may afford little protection. For some ways (other than patents) in which electronics firms protect their products from imitation, see *Science*, November 24, 1978, pp. 848–49. Also see Scherer, *ibid.* (cited in footnote 28).

Although the median increase in imitation cost due to patents (outside the drug industry) is rather small, the increase in some cases is very substantial. Outside drugs, patents increased imitation cost by 100 percent or more in about one-quarter of the cases. This helps to explain our findings in section 10 below.

9. Imitation Costs, Entry, and Concentration

We turn now from the determinants of imitation costs to their effects on entry and concentration. Holding constant the discounted profit (gross of the imitation cost) that the imitator expects to earn by imitating a new product, the new product is more likely to be imitated if the imitation cost is small. To discourage entry, the innovator may adopt pricing (and other) policies to reduce the imitator's expected discounted gross profit if the imitation cost is low. Taking this into account, is it still true that the probability of entry is inversely related to the size of the imitation cost? To find out, we determined whether each product in the sample was imitated within four years[30] after it was first introduced. (Innovations that had been on the market less than four years and unsuccessful innovations clearly had to be omitted.) Then we carried out a logit analysis to determine whether C—the ratio of imitation cost to innovation cost—influences the probability that entry of this sort occurred within four years. Letting P be the probability that such entry did *not* occur, we found that

$$(7.3) \qquad ln\left(\frac{P}{1-P}\right) = -3.10 + 3.92C.$$
$$\qquad\qquad\qquad\qquad (2.04) \quad (1.97)$$

Thus, imitation cost seems to be related in the expected way to whether or not entry occurs.[31]

Imitation cost also may affect an industry's level of concentration. We would expect an industry's concentration level to be relatively low if its members' products and processes can be imitated easily and cheaply.[32]

30. In the bulk of the cases, the new product could have been imitated in two years or less even if the imitator carried out the project at the most leisurely pace. In practically all cases it could be imitated in three years or less. Thus, four years was plenty of time for an imitator to enter.

31. Note that a one-tailed test is appropriate here. The methods used to estimate this equation are described in J. Berkson, "A Statistically Precise and Relatively Simple Method of Estimating the Bioassay with Quantal Response, Based on the Logistic Function," *Journal of the American Statistical Association* 48 (1953):565–99. For an analysis of entry rates in the chemical industry that estimates the effects of certain kinds of imitation costs, see Mansfield *et al.*, *The Production and Application of New Industrial Technology, op. cit.* (cited in Chapter 1, footnote 5), pp. 119–122. For some evidence that imitation costs influence the gap between social and private rates of return from innovations, see E. Mansfield *et al.*, "Social and Private Rates of Return from Industrial Innovations," *Quarterly Journal of Economics* 91 (1977):221–40.

32. In their simulation model, Nelson and Winter use a hypothetical "ease of imitation factor" which is defined differently from our C_i, but it seems clear that this factor is closely related to what our C_i measures. See their "Forces Generating and Limiting Concentration under Schumpeterean Competition," *Bell Journal of Economics* 9 (1978):524–48.

For each of the sixteen detailed industries included in the sample, we calculated the mean value of C_i, the mean for the jth industry being denoted by $\bar{C}_j$. Then to estimate the relationship between the mean imitation cost and the concentration level, we regressed each industry's four-firm concentration ratio, K_j, on $\bar{C}_j$, the result being[33]

$$(7.4) \qquad\qquad K_j = 6.22 + 61.5\bar{C}_j.$$
$$(8.80) \quad (12.8)$$

This finding, which seems to be the first empirical evidence regarding the relationship between the ease of imitation and the level of concentration, is entirely consistent with this hypothesis. Given the large number of factors influencing an industry's concentration level, it is interesting that this relationship is relatively close ($r^2 = .60$). Differences among industries in the technology transfer process (including transfers that are both voluntary and involuntary from the point of view of the innovator) may be able to explain much more of the interindustry variation in concentration levels than is generally recognized.

10. Patents and the Rate of Innovation

Finally, we turn to one of the most important and controversial questions concerning the patent system: what proportion of innovations would be delayed or not introduced at all if they could not be patented? To shed light on this question, we asked each innovating firm whether it would have introduced each of its patented innovations in our sample if patent protection had not been available. Although answers to such questions have obvious limitations and must be treated with caution, they should shed some light on this topic, about which so little is known.[34] According to the firms, about one-half of the patented innovations in our sample would not have been introduced without patent protection. The bulk of these innovations occurred in the drug industry. Excluding drug innovations, the lack of patent protection would have affected less than one-fourth of the patented innovations in our sample.

Taylor and Silberston have provided estimates of the proportion of R and D expenditures in various U.K. industries that would not have

33. To prevent confusion, note that each concentration ratio refers to all the detailed industry's products, not just the new products in our sample. In effect, we treat the products in our sample in each detailed industry as a sample of all the detailed industry's products, and use $\bar{C}_j$ as a measure of the ease with which all the latter products can be imitated.

34. These data could be obtained for thirty-one innovations. Some innovations could not be included because they were unpatentable or because no reliable information could be obtained from the firm. For a good discussion of what currently is known about this topic, see Scherer, op. cit. (cited in footnote 28) and Taylor and Silberston, op. cit.

been carried out without patent protection. Applying the industry weights in our sample to their data, one would expect that about 36 percent of the R and D expenditures of the firms in our sample would not have been carried out under these circumstances, if U.S. and British firms are alike in this regard. This is somewhat less[35] than the proportion of innovations that, according to the firms, would not have been introduced without patent protection.[36]

One important reason why patents frequently are not regarded as crucial is that they often have only a limited effect on the rate of entry. For about half of the innovations, the firms felt that patents had delayed the entry of imitators by less than a few months.[37] Although patents generally increased the imitation costs, they did not increase the costs enough in these cases to have an appreciable effect on the rate of entry. But although patent protection seems to have only a limited effect on entry in about half of the cases, it seems to have a very important effect in a minority of them. For about 15 percent of the innovations, patent protection was estimated to have delayed the time when the first imitator entered the market by four years or more.

According to many economists, patent protection tends to be more important to smaller firms than to larger ones. However, although this proposition seems reasonable, the existing evidence on this score is weak and sometimes contradictory.[38] To test this proposition, we carried out a logit analysis to see whether π, the probability that an innovation would have been introduced without patent protection, is related to S, the 1976 sales of the firm that carried out the innovation. Since practically none of the drug innovations would have been introduced without patent protection, such innovations were excluded.[39] The results are:

35. This difference is statistically significant at the .05 probability level.

36. There is no reason to expect the proportion of R and D expenditures that would not be carried out without patent protection to equal the proportion of innovations that would not be carried out under these circumstances. If the innovations that would not be carried out tend to be less R-and-D–intensive than those that would be carried out under these circumstances, the former proportion would be less than the latter percentage. Also, there is no reason to expect that American and British firms would behave in exactly the same way. Thus, if patents are regarded as more important in the United States than in the United Kingdom, this would result in the sort of difference we observe. In addition, there are sampling errors in both our estimates and those of Taylor and Silberston.

37. Each innovating firm was asked to estimate how much sooner the first imitating firm would have entered the market if the new product had not had patent protection. The resulting estimates are rough, but, according to the firms, they are reasonably accurate. They seem consistent with the findings in the first paragraph of this section. Outside drugs, patents were estimated to delay the entry of imitators by two years or more in 25 percent of the cases.

38. See Scherer, *op. cit.* (cited in footnote 28).

39. Had the drug innovations been included, a bias might have resulted, since, as noted above, patents are regarded as particularly important in the drug industry. Thus,

(7.5) $$ln\left(\frac{\pi}{1-\pi}\right) = 0.878 + 0.00012S.$$
$$(1.03)\quad(0.030)$$

Since the coefficient of S is far from statistically significant, there is no evidence that patent protection was more likely to be deemed essential for innovations carried out by smaller firms than for those carried out by larger ones.[40]

11. Summary

One of the major instruments of national policy regarding technology is the patent system. Over the years, there have been many proposed changes in the patent system. Some have proposed that the length of the patent be varied with the importance of the invention. Others have suggested a system of compulsory licensing or a system of government rewards to inventors. It seems unlikely that fundamental changes of this sort will be made. In December 1980, Congress passed some changes in the patent laws. Any person now can seek a reexamination by the Patent and Trademark Office of a patent. This was expected to promote the resolution of questions concerning the validity of issued patents without long and costly litigation. Also, an attempt was made to move toward a uniform patent policy toward inventions made with federal support.

The number of patents is sometimes used as a crude index of the rate of invention in a given field during a particular period of time. Used in this way, patent statistics have important disadvantages, since the average importance of the patents granted at one time and place may differ from those granted at another time and place, and the proportion of total inventions that are patented may vary considerably. Nonetheless, it is worth noting that the number of patents (by year of application) has declined during the 1970s in practically all of the product fields for which data are available. Thus, for what they may be worth, patent statistics seem to be quite consistent with the view that the rate of invention has been slowing down in the United States.

Based on a sample of forty-eight product innovations, the average ratio of imitation cost to innovation cost was about 0.65, and the average ratio of imitation time to innovation time was about 0.70. The ratio of imitation cost to innovation cost seems to be inversely related to the pro-

---

since the drug firms were larger than the average in the sample, this might have resulted in a spurious inverse relationship between π and S. S is measured in billions of dollars.

40. These results may be due in part to the fact that practically all of the innovating firms in the sample are quite large. If a larger number of very small firms were included, the results might be more in line with the proposition in the text.

portion of the product's innovation cost that goes for research. Also, this ratio tends to be large if the product is a new drug where the FDA requires that an imitative product must be tested in much the same way as the innovative product with which it competes. Further, this ratio tends to be small if the product consists of a new use of an existing material and if some firm other than the innovator has patents on this material. These three independent variables can explain about half of the variation in this ratio.

A product's imitation cost affects whether or not entry occurs in the market for the product. A logit analysis was carried out to determine whether the ratio of imitation cost to innovation cost influences the probability that entry of this sort will occur within four years of the product's first introduction. The results seem to confirm its influence. Imitation cost may also affect an industry's level of concentration. An industry's concentration level would be expected to be relatively low if its members' products and processes can be imitated easily and cheaply. Our findings, which seem to be the first empirical evidence on this score, seem to be in accord with this hypothesis.

Although patented innovations are frequently imitated within a few years of their initial introduction, patents do tend to increase imitation costs. The median estimated increase in the ratio of imitation cost to innovation cost was 11 percent. In the ethical drug industry, patents had a bigger impact on imitation costs than in the other industries. According to the relevant firms, about one-half of the patented innovations in our sample would not have been introduced without patent protection. The bulk of these innovations occurred in the drug industry. Excluding drug innovations, the lack of patent protection would have affected less than one-fourth of the innovations in our sample. One reason why patents are not regarded as crucial is that they often have only a limited effect on the rate of entry.

8 ENGINEERING EMPLOYMENT

BY AMERICAN FIRMS

1. Introduction

The nation's engineering and scientific labor force is a major resource permitting and encouraging rapid technological change and the diffusion of new products and processes. Policy makers must continually make decisions that depend, explicitly or implicitly, on forecasts of the market for engineers and scientists, and on the characteristics of the present and future engineering and scientific labor force. Yet, as many economists have pointed out,[1] very little is known about the accuracy of various techniques used to forecast engineering and scientific employment or about many important characteristics of the engineering and scientific labor force.

This chapter has three objectives. First, we describe the changes in the employment of engineers and scientists in the United States since World War II, and summarize government studies of the projected supply and utilization of engineers and scientists. Second, we present some findings regarding the accuracy of estimates made by manufacturing firms of their own engineering employment, and suggest a simple model that may be useful in improving their accuracy. Third, we present some findings regarding the age distribution, educational distribution, and earnings distribution of engineers employed by American manufacturing firms.

---

1. See G. Cain, R. Freeman, and W. L. Hansen, *Labor Markets Analysis of Engineers and Technical Workers* (Baltimore: Johns Hopkins, 1973); W. L. Hansen, "The Economics of Scientific and Engineering Manpower," *Journal of Human Resources* 2 (1967):191–215; and J. O'Connell, "The Labor Market for Engineers: An Alternative Methodology," *Journal of Human Resources* 7 (1972):71–86.

2. Scientific and Engineering Employment in the Postwar Period

Since World War II, there have been three quite distinct periods with regard to the employment of engineers and scientists. The first period, from about 1950 to 1963, was marked by rapid growth of jobs for engineers and scientists. As shown in Table 8.1, the employment of engineers and scientists grew by over 6 percent per year, which was far in excess of the rate of growth of total nonfarm employment. In part, this rapid increase was due to increases in defense activities and in the space program. During this period, there were many complaints of a shortage of engineers.[2]

TABLE 8.1

Average Annual Percentage Change in Scientific, Engineering, and Total Nonfarm Employment, 1950–63, 1963–70, and 1970–76

TYPE OF EMPLOYMENT	1950–63	1963–70	1970–76
	(*Percentages*)		
Scientists[a]	7.0	4.8	4.1
Engineers	6.5	2.5	0.4
Scientists and engineers	6.6	3.2	1.5
Nonfarm wage and salary workers	1.7	3.3	1.9

Source: National Science Foundation, *Science Indicators, 1978* (Washington, D.C.: Government Printing Office, 1979).
[a]Excludes psychologists, social scientists, and computer specialists, for which comparable data are not available.

The second period, from about 1963 to 1970, saw the employment of engineers and scientists grow at about the same rate as total nonfarm employment. The employment of scientists grew more rapidly than the employment of engineers, because there was a relatively rapid increase in college enrollments and research programs. The relatively slow rate of increase of engineering employment reflected cutbacks in defense programs and space exploration, among other things.

The third period, from about 1970 to 1976, was marked by a very slow growth of scientific and engineering employment. Whereas total

2. See Cain, Freeman, and Hansen, *ibid.;* and Hansen, *ibid.* Much of sections 2–4 is taken from Edwin Mansfield, "Education, R and D, and Productivity Growth," a paper commissioned in 1981 by the National Institute of Education.

nonfarm employment grew by 1.9 percent per year, the employment of engineers and scientists grew by 1.5 percent per year. (Indeed, between 1970 and 1972, there was a decline of 20,000 in engineering employment.) In considerable part, this was due to a slower growth (or curtailment) of college enrollment, R and D expenditures, and defense activities—particularly in aircraft and related products.

Unemployment rates for scientists and engineers have tended to be very low. During the 1960s, the unemployment rate for these workers was below one percent. But in 1971, due partly to the cutbacks in defense spending and some R and D programs, the unemployment rate for scientists and engineers rose to about 3 percent. By 1973, it fell below 1 percent once again. However, in 1975, the unemployment rate for engineers increased to 2.6 percent, due to the recession.

Most engineers and scientists are employed by industry. Over 1 million were employed in the industrial sector in the mid-1970s, as compared with about 300,000 in universities and colleges, and about 200,000 in the federal government. Table 8.2 shows the allocation of industry's labor force among various work activities. About 37 percent of the scientists and 26 percent of the engineers are involved in R and D or R and D management. However, this does not mean that the others do not play an important role in the process by which technology is developed and applied. The interface between R and D and the rest of the firm is of fundamental importance in determining the rate of innovation, as Freeman, Mansfield and Wagner, and others have indicated.[3] Production engineers, sales engineers, and other non–R and D engineers and scientists play a significant part in the innovation process.

Given the slowdown in the demand for engineers, it is not surprising that the percentage of bachelor's (and first professional) degrees awarded in engineering declined continually and significantly between 1960 and 1975. In 1960, engineering degrees were 10 percent of the total; in 1975, they were 4 percent of the total. The percentage of bachelor's (and first professional) degrees in the physical and environmental sciences fell from 4 percent in 1960 to 2 percent in 1975. (In contrast, the percentage in the social sciences increased from 8 percent in 1960 to 14 percent in 1975.) Turning from undergraduates to graduate students, enrollments for advanced degrees in science and engineering decreased from 38 percent of all advanced degree enrollment in 1960 to 25 percent in 1975.

3. Freeman, *The Economics of Industrial Innovation, op. cit.* (cited in Chapter 2, footnote 34); and Edwin Mansfield and Samuel Wagner, "Organizational and Strategic Factors Associated with Probabilities of Success in Industrial R and D," *Journal of Business* 48 (1975):179–98.

TABLE 8.2

*Percentage Distribution of Industry's Scientific and
Engineering Labor Force, by Primary Work Activity, 1974*

PRIMARY WORK ACTIVITY	SCIENTISTS	ENGINEERS	SCIENTISTS AND ENGINEERS
	(*Percentages*)		
R and D and R and D management	37	26	29
Management of non–R and D activities	15	20	19
Production and inspection	13	17	16
Design	1	18	14
Computer applications	19	2	6
Other activities	15	17	16
Totals	100	100	100

Source: National Science Foundation, *Science Indicators, 1976* (Washington, D.C.: Government Printing Office, 1977).

3. Projected Supply and Utilization of Scientists and Engineers

Government agencies—in particular, the Bureau of Labor Statistics and the National Science Foundation—have made projections of the supply and utilization of scientists and engineers in the 1980s. In its 1979 Annual Report, the National Science Board reviewed some of these projections. The Bureau of Labor Statistics is quoted as saying that there would be an ample supply of scientists through the mid-1980s. For engineers, demand and supply were expected to be roughly in balance, and engineering graduates were expected to encounter good employment opportunities through the mid-1980s.

Turning to individual fields of science, there seemed to be considerable variation in the outlook. In geology, there appeared to be favorable employment opportunities, due in part to increasing exploration for oil and other minerals. If present trends continue, a shortage of geophysicists seemed quite possible. Favorable opportunities were projected for chemists and physicists in the nonacademic sector. About three-fourths of chemists' total employment was expected to be in industry. In physics, the generally favorable prospects reflected an anticipated cutback in the supply of physicists, not an appreciable increase in the demand for them. In astronomy and mathematics, the situation seemed likely to be less rosy. The number of degrees granted in astronomy was expected to continue to exceed job openings. Mathematicians were expected to face keen competition for jobs.

In the social and life sciences, there was also appreciable variation among fields. Economists with advanced degrees were expected to have favorable job opportunities in nonacademic work. Life scientists with advanced degrees also were expected to have good opportunities. But those with training in anthropology and sociology seemed likely to encounter keen competition for jobs. For psychologists with advanced degrees, job opportunities appeared greatest for people specializing in applied areas such as industrial psychology and clinical counseling.[4]

Both the National Science Foundation and the Bureau of Labor Statistics made projections of the supply and utilization of doctoral scientists and engineers in the mid-1980s. Depending on the model used, the number of students expected to receive science and engineering doctorates (corrected for international migration) was 185,000 (the Bureau of Labor Statistics' projection for 1976–85) to 210,000 (the National Science Foundation's projection for 1977–87). Based on these projections and estimates of attrition, the labor force of doctoral scientists and engineers was expected to be about 415,000 in the mid-1980s. Of these people, about 345,000 were expected to be engaged in scientific and engineering activities.

What would happen to the remaining 70,000 doctoral scientists and engineers? Clearly, there was a very small chance that they would be unemployed. Instead, they were expected to move into other fields. Among mathematicians and social scientists, the situation looked particularly bleak. In mathematics, it was expected that 21 to 30 percent of mathematics Ph.D.s would be working outside science and engineering. In social science, the proportion was expected to be 19 to 27 percent.[5] (On the other hand, by late 1981 there were many warnings from leading engineering schools of a shortage of doctoral engineers. According to Paul Gray, president of Massachusetts Institute of Technology, there was a need for more young people preparing for faculty careers in engineering and some areas of applied science.)

4. Accuracy of Firms' Forecasts of Engineering Employment

Policy makers in government, universities, and business must make decisions that depend, explicitly or implicitly, on forecasts of the number of engineers employed in various sectors of the economy at various points in time. For example, in evaluating the adequacy of existing engineering

4. *Science Indicators, 1978, op. cit.* (cited in Chapter 7, footnote 15). Also, see *Science Indicators, 1980* (Washington, D.C.: National Science Foundation, 1981).
    5. *Ibid.*

manpower, public policy makers must try to forecast how many engineers will be employed in the private sector. Many groups, including the National Science Foundation and the Bureau of Labor Statistics, have made forecasts of this sort for decades.[6] Although such forecasts sometimes are based on a collection of forecasts made by firms of their own engineering employment, very little is known concerning the accuracy of firms' forecasts of this kind.

In this section, we present data concerning the accuracy of such forecasts and suggest a simple model that may be useful in improving their accuracy. Very detailed data were obtained from a well-known engineering association which has collected such forecasts from firms for many years. For fifty-four firms in the aerospace, electronics, chemical, and petroleum industries, comparisons were made of each firm's forecasted engineering employment with its actual engineering employment during 1957 to 1976. Since data were obtained concerning a number of forecasts of each firm, the accuracy of 218 such forecasts could be evaluated.[7]

The results indicate that the forecasting errors for individual firms in the aerospace industry have been large, as can be seen in Table 8.3. For example, even when firms have forecasted only six months ahead,

6. For example, see *ibid.* and National Science Foundation, *Projections of Science and Engineering Doctorate Supply and Utilization, 1982 and 1987* (Washington, D.C.: Government Printing Office, 1979), as well as H. Folk, *The Shortage of Scientists and Engineers* (Lexington: Heath Lexington, 1970).

7. These four industries were chosen because they employ almost 60 percent of all engineers in manufacturing. The firms that were included tend to be among the largest in these industries. All firms are included for which usable data were obtained by the engineering association. Each firm forecasted what its engineering employment would be at the end of about six months, two years, five years, and ten years. Subsequently, each such forecast was compared with the firm's actual engineering employment then. (In some cases, the firm provided a high and low forecast, in which case the mean was used.) Both the forecast and the actual figure were reported to the engineering association from which we received the data.

In a relatively few cases, the actual figure had to be approximated by interpolation between figures given for somewhat different dates. Also, in a relatively few cases, because the firm did not include nongraduate engineers in its figures, it was necessary to assume that the ratio of nongraduate to graduate engineers equaled what it was in an earlier period. These adjustments have no effect at all on the results pertaining to the six-month forecasts, since none of the observations was adjusted. They almost always have very little effect on the results pertaining to the two-year or five-year forecasts. But for the ten-year forecasts, the errors for the unadjusted observations tended to be smaller than for all observations. Thus, our results may overstate the forecast errors for the ten-year forecasts.

An engineer is defined here as someone who has attained engineering status through company training or work experience as well as through a degree in engineering. Such nongraduates comprise a small percentage of total engineering employment. For a good discussion of various definitions of an engineer, see Cain, Freeman, and Hansen, *op. cit.*

TABLE 8.3

*Mean Percentage Error in a Firm's Forecast of
Its Engineering Employment, Fifty-four Firms, 1957–76*

	FORECASTING INTERVAL[a] (YEARS)			
INDUSTRY	0.5	2	5	10
Aerospace	10.3	15.9	41.2	88.7
Electronics	4.6	12.4	15.4	26.5
Chemicals	3.2	5.7	17.3	22.0
Petroleum	2.8	5.5	13.1	9.4

[a] The forecasting interval is the length of time between the date when the forecast is made and the date to which it applies.

the mean percentage error was about 10 percent. In the electronics, chemical, and petroleum industries, the forecasting errors for individual firms have been much less, although the mean percentage error for two-year forecasts in the electronics industry was about 12 percent. The relatively large forecasting errors in the aerospace industry (and to a lesser extent, the electronics industry) seem to be due to its heavy dependence on government defense and space programs, which were volatile and hard to predict.

As might be expected if there is no major bias in the forecasts, although the forecasting errors for individual firms are substantial, they tend to be smaller when we consider the total engineering employment for all firms in the sample. On the average, the six-month forecasts were in error by about 2 percent, the two-year forecasts were in error by about 1 percent, and the five-year forecasts were in error by about 3 percent.[8] The fact that there was so little bias in the forecasts is encouraging since, for many purposes, the principal aim is to forecast total engineering employment in some sector of the economy, not the engineering employment of a particular firm.

Studies of forecast errors often indicate that, when the actual quantity is relatively high, the forecast is too low, and that, when the actual quantity is relatively low, the forecast is too high (as illustrated by line *AB* in Figure 8.1). This was found to be true by Tull in his study of the accuracy of the forecasts of new product sales, and by Beardsley and Mansfield in their study of the accuracy of forecasts of the profitability of new products and processes.[9] However, our data indicate that this is

8. The ten-year forecasts showed an average upward bias of about 27 percent, due no doubt in part to the extrapolation of the relatively high rates of growth in engineering employment during the late 1950s and early 1960s to later periods. Also see footnote 7.

9. See D. Tull, "The Relationship of Actual and Predicted Sales and Profits in New Product Introductions," *Journal of Business* (1967); and G. Beardsley and E. Mansfield, "A

FIGURE 8.1

*Alternative Possible Relationships between Actual
and Forecasted Quantities*

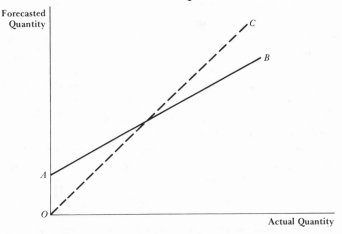

not the case for engineering employment. On the contrary, holding constant the length of the forecasting interval, the average relationship between forecasted and actual employment by firms can be approximated by a straight line through the origin, as represented by line OC in Figure 8.1. Based on regression analysis, we found no statistically significant indication that the intercept of the actual versus forecasted line is nonzero.[10]

Economists sometimes construct models in which it is hypothesized that firms at each point in time have a desired employment level for a particular kind of labor, and that they set their actual employment level for this kind of labor so as to move part way toward this desired employment level. Thus, in the case of engineers, firms are continually adjusting their employment toward the level that they would regard as optimal if changes in employment levels could be made instantaneously and if the inefficiencies involved in too rapid a change in engineering employment could be avoided. If $E_i(t)$ is the ith firm's engineering employment at time t, and if $\bar{E}_i(t + 1)$ is its optimal or desired employment one year hence, then $E_i(t + 1)$ can be represented as

$$(8.1) \qquad E_i(t + 1) = E_i(t) + \theta_i(t) \, [\bar{E}_i(t + 1) - E_i(t)].$$

Note on the Accuracy of Industrial Forecasts of the Profitability of New Processes and Products," *op. cit.* (cited in Chapter 6, footnote 10).

10. Holding industry and forecasting interval constant, when a firm's forecasted employment is regressed on its actual employment, the intercept of the regression (in twelve out of thirteen cases) does not differ significantly from zero (at the .05 level).

In other words, $\theta_i(t)$ is the proportion of the way that the ith firm's engineering employment moves toward the desired level between time t and time $t + 1$.

If one can estimate $\theta_i(t)$, equation (8.1) can be used to forecast $E_i(t + 1)$, since data can be obtained at time t regarding $E_i(t)$ and $\bar{E}_i(t + 1)$. The engineering association collected data concerning $\bar{E}_i(t + 1)$ for various times between 1957 and 1968, so we were able to obtain direct estimates of $\theta_i(t)$ for seven major chemical firms and six major petroleum firms during this period. These were all of the firms for which appropriate data were available.[11] The mean value of $\theta_i(t)$ is very similar for the two industries: it is 0.73 in chemicals and 0.72 in petroleum. These results are quite similar to those of Freeman, although his estimates of the rate of adjustment are based on quite different kinds of data.[12]

To explain differences among time periods and firms in the value of $\theta_i(t)$, it seems reasonable to assume that

$$(8.2) \qquad \theta_i(t) = \phi_0 + \phi_1 D_i(t) + \phi_2 I_i(t) + u_i(t),$$

where $D_i(t)$ is the desired proportional increase in engineering employment between time t and time $t + 1$, $I_i(t)$ is the ratio of the ith firm's profits in time t to those in time $t - 1$, and $u_i(t)$ is a random error term.[13] A priori, we would expect ϕ_1 to be negative since, if attaining its desired employment level means that the firm must increase its employment by a relatively large percentage, this firm will move a relatively small proportion of the way toward this desired level because of the costs of rapid change in employment levels. Similarly, we would expect ϕ_2 to be positive because relatively large increases in profits will influence firms' expectations and make them bolder in moving toward desired employment levels.

To see how well the hypothesized model in equation (8.2) fits the data, we obtained least-squares estimates of ϕ_0, ϕ_1, and ϕ_2, as shown in Table 8.4. The results show that each of the regression coefficients has

11. The engineering association, cited above, obtained data during this period regarding each firm's desired and actual levels of engineering employment. Such data were not obtained for previous or subsequent years. The definition of the desired level was not exactly the same as that given above, but it was regarded by officials of the association as being reasonably close. It called for the number of engineers desired to meet the firm's goals and commitments. All firms in these industries were included where $\bar{E}_i(t + 1) > E_i(t)$. See footnote 13.

12. See R. Freeman, "Scientists and Engineers in the Industrial Economy," unpublished National Science Foundation report, 1977.

13. This model assumes that $\bar{E}_i(t + 1) > E_i(t)$, which was the typical case during 1957–68. The data regarding $I_i(t)$ come from *Moody's*. For the chemical firms, $I_i(t)$ is the ith firm's net income in year $t + 1$ divided by its net income in year t; for the petroleum firms, it is the firm's net income in year t divided by its net income in year $t - 1$. $D_i(t)$ equals $[\bar{E}_i(t + 1) - E_i(t)] \div E_i(t)$.

TABLE 8.4

Estimated Regression Coefficients,[a] *Equation (8.2)*

INDUSTRY	INDEPENDENT VARIABLES			
	ϕ_0	$D_i(t)$	$I_i(t)$	$\bar{R}^2$
Chemical	0.32 (0.95)	−4.61 (5.8)	0.74 (2.15)	0.78
Petroleum	−1.21 (1.2)	−6.55 (3.1)	2.33 (2.35)	0.51

[a] The *t*-statistic is shown in parentheses below each regression coefficient.

the expected sign and is statistically significant. This model explains over three-quarters of the variation in $\theta_i(t)$ in chemicals and about one-half of such variation in petroleum.[14] Using the least-squares estimates of ϕ_0, ϕ_1, and ϕ_2, one can estimate $\theta_i(t)$ for each firm on the basis of its values of $D_i(t)$ and $I_i(t)$. Inserting this estimate of $\theta_i(t)$ into equation (8.1), one can forecast $E_i(t+1)$. Based on the data for these firms, the resulting forecasts are appreciably better than those of the firms themselves.[15]

In view of the importance of technical manpower in promoting technological change in both civilian and military fields, both government agencies and private organizations must continually assess, as best they can, the future state of the labor market for many kinds of professional and technical personnel. Our results pertain entirely to engineers, but the approach adopted above may be applicable to other types of personnel as well. Better forecasts may result from the application of this simple sort of model. At the same time, our findings are tentative in many respects, even for engineers.[16] This seems to be one of the first reason-

14. Mansfield found that a similar model worked quite well for R and D expenditures in these industries. See his *Industrial Research and Technological Innovation, op. cit.* (cited in Chapter 1, footnote 1). Also, see Edwin Mansfield, "How Economists see R and D," *Harvard Business Review* (November 1981).

15. The mean percentage error for the chemical firms is about 0.25 percentage points and for the petroleum firms is about 1.25 percentage points, which is considerably less than in Table 8.3. Since the estimates of the ϕ's are based on the same data, these results are essentially a measure of goodness of fit. The next step might be to try out a model of this sort over a reasonably long period of time, and see how well it works and how it can be improved. Obviously, the present results are only a first step.

16. Although we are fortunate to have such a large sample in Table 8.3 (over two hundred forecasts), the sample is by no means ideal. For example, many of the forecasts relate to a considerable number of years ago. It would be desirable to have more information on changes over time in the accuracy of firms' forecasts, but the data obtained from the engineering association cannot support an adequate study of this matter because a number of relevant variables cannot be held constant and for other reasons. For additional limitations of this study, see footnotes 7, 11, 13, and 15.

ably comprehensive studies of firms' forecasts of engineering employment. Much more needs to be done.

5. Age Distribution of Engineers

Policy makers must be concerned with the characteristics of the engineering labor force, as well as with the number of engineers employed. One important characteristic of the nation's engineers is their age. Some studies suggest that engineers, particularly those engaged in research and development, tend to experience a reduction in creativity after the age of thirty-five or forty, due in part to obsolescence of knowledge.[17] Given the slowdown in the rate of growth of the engineering labor force in the past twenty years, one would expect that the proportion of engineers under thirty-five or forty has declined. To see how big this decline has been, we obtained very detailed data concerning the age distribution of engineers employed by six major firms in the aerospace, chemical, and petroleum industries during 1960–74. The results in Table 8.5 show that the percentage of engineers under thirty-four years (and under forty-three years) decreased by over twenty percentage points in these firms during 1960–74.[18]

TABLE 8.5

*Percent of Engineers Less Than Thirty-four Years Old
and Less Than Forty-three Years Old,
Four Industries, 1960–74*

	PERCENT UNDER THIRTY-FOUR YEARS			PERCENT UNDER FORTY-THREE YEARS		
		BASED ON REGRESSIONS			BASED ON REGRESSIONS	
YEAR	SIX MAJOR FIRMS	AEROSPACE AND ELECTRONICS	CHEMICAL AND PETROLEUM	SIX MAJOR FIRMS	AEROSPACE AND ELECTRONICS	CHEMICAL AND PETROLEUM
1974	30	29	21	56	66	62
1970	36	—	—	67	—	—
1966	38	—	—	72	—	—
1960	51	65	36	78	92	81

17. For example, see W. Decker and C. Van Atta, "Controlling Staff Age and Flexible Retirement Plans," *Research Management* 16 (1973):16–21; and D. Pelz and F. Andrews, *Scientists in Organizations* (New York: Wiley, 1966).

18. These data were obtained from the firms by the engineering association cited in previous sections. We use thirty-four years and forty-three years because they are cutoff points in the original data. These firms were included because they reported data for all four years in Table 8.5. In this and subsequent sections, an engineer is defined as someone trained as an engineer, even if he or she is not working as an engineer.

Utilizing data for eighteen firms in the aerospace, electronics, chemical, and petroleum industries,[19] we also regressed the percent of a firm's engineers under thirty-four years (or under forty-three years) in a given year on the firm's sales, profits, ratio of engineers to total employment, ratio of scientists to engineers, and time. This regression can be used to estimate the change between 1960 and 1974 in this percentage when all of these variables (other than time) are held constant at their mean values. The results, shown in Table 8.5, confirm the finding of our six-firm sample that the engineering labor force in these industries has aged considerably. If, as some claim, the rate of innovation in the United States has been slowing down, this "graying" of industry's engineers may be partly responsible.[20]

6. Educational Distribution of Engineers

Another important characteristic of the engineering labor force is its educational distribution. In some firms and time periods, the percentage of engineers with advanced degrees is much higher than in others. In part, this is because some firms, but not others, find it profitable to carry out the highly specialized and relatively fundamental technical work for which engineers with advanced degrees are trained. It is interesting to see whether relatively large firms, which (on the average) devote a bigger percentage of their technical resources to basic research than smaller firms,[21] tend to have a larger percentage of engineers with Ph.D.s or master's degrees than smaller firms. Holding size of firm constant, do manufacturing firms where engineers constitute a relatively large percentage of the firm's total employment tend to have a relatively high percentage of engineers with Ph.D.s or master's degrees?[22] Holding other factors constant, do manufacturing firms that hire a relatively large

19. These firms were the ones that responded relatively consistently and in detail to the engineering association. The regressions are based on data for the even years between 1960 and 1974 (except that, for reasons indicated in footnote 25, 1972 had to be omitted). A further description of the sample is provided in section 6 and footnote 25.

20. The Organization for Economic Cooperation and Development has pointed out that, among government and academic scientists and engineers, this "graying" of the labor force is evident in many countries, but among industrial scientists and engineers, relatively little seems to be known on this score. See *Trends in Industrial R and D in Selected Member Countries, 1967–75* (Paris: Organization for Economic Cooperation and Development, 1979).

21. See Mansfield, Rapoport, Schnee, Wagner, and Hamburger, *Research and Innovation in the Modern Corporation, op. cit.* (cited in Chapter 1, footnote 6).

22. With firm size held constant, increases in the number of engineers employed may result in greater specialization of engineers and more need for engineers with advanced degrees.

number of scientists (per engineer) tend to have a relatively high percentage of engineers with Ph.D.s or master's degrees?[23]

The educational distribution of a firm's engineers is determined by relative salary levels. For example, if the salary of engineers with master's degrees rises relative to the salary of engineers with bachelor's degrees, we would expect firms to substitute B.S. engineers for those with master's degrees. Further, it is sometimes argued that the extent to which a firm is willing and able to carry out the specialized work requiring engineers with advanced degrees is dependent on the firm's profitability. According to Nason, Steger, and Manners, some firms have cut back considerably on their basic research expenditures because of reductions in the firms' profitability.[24]

To test whether these factors affect the educational distribution of a firm's engineering labor force, data were obtained regarding seven firms in the aerospace and electronics industries, eight petroleum firms, and three chemical firms. (Almost all of these firms are among the five hundred largest manufacturing firms in the United States.) The data pertain to the even years from 1958 to 1974 (with 1972 excepted). Using these data, we obtained least-squares estimates of the α's in the following equations:

(8.3a) $\quad B_{it} = \alpha_{01} + \alpha_{11}S_{it} + \alpha_{21}\pi_{it} + \alpha_{31}G_{it} + \alpha_{41}H_{it} + \alpha_{51}W_{it} + z_{1it}$

(8.3b) $\quad M_{it} = \alpha_{02} + \alpha_{12}S_{it} + \alpha_{22}\pi_{it} + \alpha_{32}G_{it} + \alpha_{42}H_{it} + \alpha_{52}W_{it} + z_{2it}$

(8.3c) $\quad P_{it} = \alpha_{03} + \alpha_{13}S_{it} + \alpha_{23}\pi_{it} + \alpha_{33}G_{it} + \alpha_{43}H_{it} + \alpha_{53}W_{it} + z_{3it},$

where B_{it} is the percent of the ith firm's engineers that have only bachelor's degrees in year t, M_{it} is the percent of the ith firm's engineers that have master's degrees (but not Ph.D.s) in year t, P_{it} is the percent of the ith firm's engineers that have Ph.D.s in year t, S_{it} is the ith firm's deflated sales in year t (expressed in billions of 1967 dollars), π_{it} is the ratio of net income to net worth for the ith firm in year t, G_{it} is the ratio of engineers to total employment in the ith firm in year t, H_{it} is the ratio of scientists to engineers in the ith firm, W_{it} is the mean salary of engineers with the education corresponding to the dependent variable divided by the unweighted mean salary for all three educational levels, and the z's are random error terms.[25] One regression was calculated for the aerospace

23. Firms that a hire a relatively large number of scientists per engineer may be engaged in a different sort of technical work than other firms, and they may have more need for engineers with advanced degrees.

24. Nason, Steger, and Manners, *Support of Basic Research by Industry, op. cit.* (cited in Chapter 6, footnote 27).

25. The data regarding B_{it}, M_{it}, P_{it}, and W_{it} were collected by the engineering association cited in previous sections. All firms were included that responded relatively consistently and in detail. This eighteen-firm sample is the same as that used in section 5. The

and electronics firms; another was calculated for the chemical and petro-leum firms.[26]

The results, shown in Table 8.6, indicate that there is a very signifi-cant tendency for the percentage of engineers with advanced degrees to be higher (and the percent with bachelor's degrees to be lower) in larger firms than in smaller ones. There is also a very significant tendency for firms with a relatively high ratio of scientists to engineers to have a high

TABLE 8.6

Estimated Regression Coefficients, Equation (8.3)

DEPENDENT VARIABLE	CONSTANT	S_{it}	π_{it}	G_{it}	H_{it}	W_{it}	$\bar{R}^2$	NUMBER OF OBSERVATIONS
			Aerospace and Electronics					
B_{it}	163.70	−7.90	19.52	−163.40	−55.76	−63.96	0.82	33
	(9.05)	(6.68)	(1.22)	(7.26)	(4.26)	(2.85)		
M_{it}	51.27	7.12	−1.24	168.90	49.30	−66.39	0.78	33
	(1.44)	(6.90)	(0.80)	(7.95)	(4.00)	(1.84)		
P_{it}	15.09	1.80	−7.31	6.04	−0.25	−11.59	0.78	33
	(2.60)	(8.00)	(2.38)	(1.39)	(0.90)	(2.47)		
			Chemical and Petroleum					
B_{it}	146.80	−3.00	−32.92	108.60	−6.50	−63.91	0.36	60
	(5.23)	(4.20)	(1.48)	(2.89)	(3.00)	(1.94)		
M_{it}	30.80	1.84	41.60	−84.17	4.97	−24.10	0.25	60
	(1.09)	(3.10)	(2.41)	(2.71)	(2.93)	(0.81)		
P_{it}	23.78	1.03	1.47	−10.18	2.52	−20.58	0.39	60
	(2.25)	(4.50)	(0.20)	(0.83)	(3.60)	(2.34)		

percentage of engineers with advanced degrees. Further, the percent of engineers with each degree level is inversely related to its salary level (relative to the others' salary levels). However, the profitability of the firm seems to have no consistent effect on the educational distribution of the firm's engineers; and whereas increases in the ratio of engineers to total employment seem to be associated with increases in the percent of engineers with advanced degrees in aerospace and electronics, the opposite tendency prevails in chemicals and petroleum.

engineering association obtained data for even years only. We had to exclude 1972 because the engineering association had discarded the questionnaires received that year. The data regarding W_{it} are industry averages. The data regarding π_{it} come from *Moody's;* the sales data are deflated by the price index for the relevant industry. Each firm's total employment comes from its annual reports.

26. Since the chemical and petroleum industries have much in common (technologi-cally and economically) and the aerospace and electronics industries have much in com-mon, we pooled them in these regressions.

7. Interfirm Differences in Discounted Engineering Earnings

Little is known about the differences among firms in the discounted lifetime earnings of their engineers. Information on this score would be a useful indicator of interfirm differences in the quality of engineers, in the types of jobs engineers perform, and in the nonmonetary benefits they receive. To obtain such information, the relationship between the median salary of engineers and their age was obtained for each of the eighteen firms included in sections 5 and 6 in each even year from 1958 to 1974 (with 1972 excepted).[27] Based on each of these relationships, the discounted earnings of engineers can be estimated, if we make the customary assumption that, on the average, an engineer's salary would vary over time in accord with this cross-section relationship.[28] Using an interest rate of 6 percent, we calculated such discounted engineering earnings for each firm and for each year for which data exist. The results indicate that, at a particular point in time, major firms in the same industry differ by 10 to 20 percent with regard to the average discounted lifetime earnings of their engineers.

What sorts of firms tend to pay relatively high discounted lifetime earnings to engineers? One relevant variable may be a firm's size. If, as some claim, the quality of the engineers employed by the larger manufacturing firms tends to be higher, on the average, than that of the smaller manufacturing firms (in the same industry), one might expect that the level of discounted earnings would be directly related to size of firm. Also, there may be nonmonetary benefits in smaller firms, such as a less structured type of organization, for which engineers may be willing to trade off income.

Holding size of firm constant, there may be a tendency for firms where engineers constitute a relatively large percentage of total employment to pay less for engineers (of a given educational level) than other firms do. As a firm expands the proportion of its work force that are engineers, it may tend to substitute engineers for nonengineers in some relatively low-paid work. Also, it may tend to substitute engineers with

27. These data were collected by the engineering association cited in previous sections. See footnote 25 for the reason why 1972 had to be excluded.
28. The roughness of this assumption is obvious. For some studies which have used and / or discussed this assumption (and its limitations), see Cain, Freeman, and Hansen, *op. cit.*; Hansen, *op. cit.*; and H. Folk, *op. cit.* Because an engineer, as defined here, includes people trained in engineering even if they are not working as engineers, these data include the earnings of people trained in engineering who become managers. Since many engineers enter management, it would be misleading for many purposes to exclude them.

more advanced degrees for those with somewhat less education. Because of such substitution, discounted engineering earnings may be inversely related to the percent of a firm's total employees that are engineers.

In addition, it seems clear that, all other things equal, the level of discounted engineering earnings generally has increased over time. Finally, it is possible that relatively profitable firms tend to have relatively high discounted engineering earnings, and that a firm's level of discounted engineering earnings is related to its ratio of scientists to engineers.

To see how well these factors can explain the observed variation in the level of discounted engineering earnings, we assume that

(8.4a) $Y_{bit} = \beta_{01} + \beta_{11}S_{it} + \beta_{21}\pi_{it} + \beta_{31}G_{it} + \beta_{41}H_{it} + \beta_{51}t + \epsilon_{1it}$

(8.4b) $Y_{mit} = \beta_{02} + \beta_{12}S_{it} + \beta_{22}\pi_{it} + \beta_{32}G_{it} + \beta_{42}H_{it} + \beta_{52}t + \epsilon_{2it}$

(8.4c) $Y_{pit} = \beta_{03} + \beta_{13}S_{it} + \beta_{32}\pi_{it} + \beta_{33}G_{it} + \beta_{43}H_{it} + \beta_{53}t + \epsilon_{3it},$

where Y_{bit} is the discounted earnings of engineers with only bachelor's degrees in the ith firm in year t, Y_{mit} is the discounted earnings of engineers with master's degrees (but not Ph.D.s) in the ith firm in year t, Y_{pit} is the discounted earnings of engineers with Ph.D.s in the ith firm in year t, and the ϵ's are random error terms. Using least-squares, we estimated the β's. One regression was computed for the chemical and petroleum firms; another was computed for the aerospace and electronics firms.[29]

The results, shown in Table 8.7, indicate that discounted engineering earnings tend to be higher in relatively large than in relatively small firms. For example, a $1-billion increase in firm sales seems to be associated with a $12,040 increase in discounted earnings of engineers with bachelor's degrees in the aerospace and electronics industries. Also, increases in the proportion of employees that are engineers generally tend to be associated with decreases in discounted earnings. For example, an increase of .10 in this proportion is associated with a $19,030 decrease in discounted earnings for engineers with bachelor's degrees in the aerospace and electronics industries. In addition, there usually is a significant upward trend in discounted earnings, as expected. No significant relationship exists between a firm's profitability and the level of its discounted engineering earnings. There is a significant relationship between a firm's ratio of scientists to engineers and its level of engineering earnings, but in chemicals and petroleum this relationship is direct,

29. As indicated in the footnotes to Table 8.7, a linear trend was not used in the aerospace and electronics industries because salary levels there did not tend to increase at a constant rate.

TABLE 8.7

Estimated Regression Coefficients, Equation (8.4)

DEPENDENT VARIABLE[a]	CONSTANT	INDEPENDENT VARIABLE					$\bar{R}^2$	NUMBER OF OBSERVATIONS
		S_{it}	π_{it}	G_{it}	H_{it}	t		
Aerospace and Electronics								
Y_{bit}	210	12.04	−28.9	−190.3	−90.3	[b]	0.82	37
	(18.8)	(4.80)	(0.79)	(3.70)	(3.00)			
Y_{mit}	230	8.40	33.0	−230.5	−65.3	[c]	0.85	27
	(15.1)	(2.06)	(0.84)	(2.52)	(2.14)			
Y_{pit}	300[d]	29.1	−0.65	−249.8	−212.4	[e]	0.73	16
	(8.5)	(3.1)	(0.007)	(1.25)	(1.77)			
Chemical and Petroleum								
Y_{bit}	−3,980	4.62	15.37	27.13	13.82	2.11	0.74	57
	(7.5)	(3.1)	(0.41)	(0.38)	(3.9)	(7.8)		
Y_{mit}	−2,873	7.59	−51.10	−128.7	29.31	1.56	0.73	44
	(4.3)	(4.3)	(1.11)	(1.56)	(6.6)	(4.6)		
Y_{pit}	−2,602	13.69	−91.54	−358.4	35.51	1.44	0.75	34
	(2.57)	(4.8)	(1.47)	(3.09)	(4.9)	(2.79)		

[a]The dependent variable is expressed in thousands of dollars.
[b]Rather than assume that the trend is linear, time dummy variables are used. The estimated coefficients (in thousands of dollars) of these variables are 21.2 (1970), 20.7 (1968), 17.7 (1966), 19.4 (1964), 9.5 (1962), −2.6 (1960), and −21.7 (1958). Each of these coefficients shows the difference between the year in question and 1974.
[c]Rather than assume that the trend is linear, time dummy variables are used. The estimated coefficients (in thousands of dollars) of these variables are 17.1 (1970), 8.5 (1968), 12.0 (1966), 19.8 (1964), 6.7 (1962), 3.2 (1960), and −13.5 (1958). Each of these coefficients shows the difference between the year in question and 1974.
[d]This is the constant for the aerospace firms. In this regression (but not the others), an industry dummy variable (equaling 1 for electronics firms and 0 for aerospace firms) is significant. The coefficient of this dummy variable is −28.3 thousands of dollars.
[e]In this case, a time trend is not statistically significant.

whereas in aerospace and electronics it is inverse. In both industry groups, these regressions seem to explain about 80 percent of the interfirm and temporal variation in discounted engineering earnings.

8. Summary

Since World War II, there has been a noteworthy reduction in the rate of increase of total employment of engineers and scientists. Whereas the employment of engineers and scientists grew much more rapidly than total nonfarm employment in 1950–63 (due partly to rapid increases in defense and space activities), it grew much less rapidly than total nonfarm employment in the 1970s (due partly to slower growth of college enrollment and R and D). Given the slowdown in the demand for scientists and engineers, it is not surprising that the percentage of bachelor's degrees awarded in engineering and the physical and environmental sciences, and the percentage of graduate students enrolled for advanced degrees in science and engineering decreased considerably between 1960 and 1975. Most engineers and scientists are employed by industry, rather than government or universities. About 37 percent of the scientists and 26 percent of the engineers are involved in R and D or R and D management.

Government agencies like the Bureau of Labor Statistics and the National Science Foundation make projections of the supply and utilization of engineers and scientists. Such projections help to guide policy makers' decisions concerning the adequacy of the nation's scientific and engineering labor force. Forecasts made by firms of their own engineering employment seem to contain relatively little bias. For individual firms, the forecasting error tends to be much greater in the aerospace and electronics industries than in the chemical and petroleum industries, due in part to the greater dependence of the former industries on the government market. According to the available data, the proportion of the way that a firm's engineering employment moves toward the desired level in one year depends on the desired percentage increase in engineering employment and the percentage change in profits. An econometric model based on this relationship may be useful in reducing forecasting errors.

Due to the slowdown in the demand for engineers, the proportion of engineers that are relatively young has declined considerably. For example, in six major firms in the aerospace, chemical, and petroleum industries, the percentage of engineers under thirty-four years (and under forty-three years) decreased by over 20 percentage points between 1960 and 1974. It is important to recognize that this "graying" of the engineering labor force has occurred in industry as well as in the universities, where the problem has received much more attention. Since some studies suggest that engineers, particularly those engaged in research and development, tend to experience a reduction in creativity after the age of thirty-five or forty, this shift in the age distribution of the engineering labor force may have tended to dampen—and may continue to dampen—the rate of innovation in the United States.

There are considerable differences among firms in the educational distribution of their engineers. The percentage of engineers with advanced degrees tends to be relatively high in larger firms, and in firms with a relatively high ratio of scientists to engineers. The percentage of engineers with each degree level is inversely related to its salary level (relative to the others' salary levels). There are also appreciable differences among firms in discounted engineering earnings. At a particular point in time, major firms in the same industry differ by 10 to 20 percent with regard to the average discounted lifetime earnings of their engineers. Holding educational level constant, discounted engineering earnings tend to be relatively high in large firms and in firms where the ratio of nonengineers to engineers is relatively high. There is no apparent relationship between a firm's profitability and its discounted engineering earnings (or the educational distribution of its engineers).

9 FEDERAL SUPPORT OF R AND D

ACTIVITIES IN THE PRIVATE SECTOR

1. Introduction

As indicated repeatedly in previous chapters, many Americans, both in the private sector and in the public sector, have been concerned about the vitality and future of civilian technology in the United States. And a variety of questions has arisen concerning the role of the federal government in stimulating civilian technology. To what extent does the federal government support research and development in the private sector? How is this support distributed among industries, universities, research centers, and other organizations? What incentives are there for private recipients to control costs or improve the efficiency of federally funded R and D activities? Why is support of this kind regarded as being in the public interest? What measurements have been made of the social benefits of additional investments in R and D, both in agriculture and industry? Is there any evidence of an underinvestment in particular types of civilian technology? What mechanisms of government support have been used in other countries, such as Japan, France, and the United Kingdom? In the United States, what are the major advantages and disadvantages associated with each of the mechanisms for federal support of private sector R and D? What are some possible approaches to improving the effectiveness of federal programs in support of R and D in the private sector?

The purpose of this chapter is to take up each of these questions. To keep the discussion to a reasonable length, we shall have to treat many of them rather cursorily. To prevent confusion, the private sector should be defined. We regard all privately owned firms and nonprofit organizations as belonging to the private sector. However, we recognize that some such firms and organizations do a heavy volume of business with the government and are so closely linked with government agencies that the distinction between the private sector and the public sector can be somewhat blurred.

2. Federal Contracts and Grants for R and D

To begin with, we must look briefly at the present extent and pattern of federal support of R and D activities in the private sector. An important part of this support is encompassed by federal contracts and grants for research and development. Total expenditures in the United States for R and D were about $60 billion in 1980, of which about $29 billion were financed by the federal government (Table 9.1.) Thus, about 48 percent of our nation's R and D expenditures were financed by the federal government in 1980, and much of this federally financed R and D was carried out by the private sector. As shown in Table 9.1, government laboratories carried out less than 30 percent of federally financed R and D. About 50 percent of federally financed R and D was carried out by industry.

Federal R and D expenditures are concentrated heavily in a relatively few areas. In 1979, as shown in Table 9.2, almost $14 billion was spent on defense R and D, and over $3 billion was spent on space R and D. Health R and D accounted for over $3 billion, and energy R and D accounted for about $2.8 billion. Other areas where significant amounts of federally financed R and D took place were environmental protection,

TABLE 9.1

*Sources of R and D Funds
and Performers of R and D, by Sector,
United States, 1980[a]*

SECTOR	FUNDS FOR R AND D	R AND D PERFORMANCE
Federal government	29,400	7,830
Industry[b]	28,710	42,250
Academic sector		
Colleges and universities	1,300	6,050
Associated FFRDCs[c]	—	2,000
Other nonprofit institutions[d]	965	2,245
Totals	60,375	60,375

Source: National Science Foundation, "National R and D Spending Expected to Reach $67 Billion in 1981," *Science Resources Studies Highlights,* May 23, 1980.
[a]Estimates made by the National Science Foundation.
[b]Includes expenditures for federally funded research and development centers (FFRDCs) administered by industry. They account for less then 5 percent of the total performance.
[c]FFRDCs administered by individual colleges and universities and by university consortia.
[d]Includes expenditures for FFRDCs administered by nonprofit organizations. They account for less than 15 percent of the total performance.

TABLE 9.2

*Federal Obligations for R and D
for Selected Functions, 1969 and 1979*

FUNCTION	1969	1979[a]
	(Millions of Dollars)	
National defense	8,356	13,833
Space	3,732	3,383
Health	1,127	3,034
Energy	435	2,827
Science and technology base	436	1,061
Environment	285	1,082
Transportation and communications	458	837
Natural resources	199	644
Food and fiber	225	543
Education	155	146

Source: Science Indicators, 1978 (Washington, D.C.: National Science Foundation, 1979).
[a] Estimates made by the National Science Foundation.

transportation, agriculture, and education. A considerable amount was spent by the federal government on the general advancement of science and technology. Despite the fact that defense and space R and D were a smaller percentage of total federally financed R and D than they were a decade before, they still constituted about 60 percent of the total.

The extent to which various federal agencies perform R and D outside government laboratories differs considerably. As shown in Table 9.3, the Department of Defense performs about one-fourth of its R and D in government laboratories; most of the remainder is performed by industrial firms. Similarly, NASA performs about one-quarter of its R and D in government laboratories; the rest is performed largely by industrial firms. On the other hand, the Department of Energy performed about half of its R and D in federally funded research and development centers (like Oak Ridge, Sandia, Brookhaven, and Los Alamos), some of which are administered by firms, some by universities. And other agencies, like the Department of Agriculture and the Department of Commerce, perform most of their R and D in their own laboratories.

There are also very substantial differences among industries in the extent to which the R and D that they perform is financed by the federal government. As shown in Table 9.4, in 1978 the federal government financed about 80 percent of the R and D in the aircraft industry and about 50 percent of the R and D in the electrical equipment industry. These are the industries where the largest share of the R and D performance is federally financed. On the other hand, in the chemical, petro-

TABLE 9.3

Federal Obligations for R and D
in Major Agencies, by Performer, Fiscal Year, 1979[a]

AGENCY	INTRAMURAL	INDUSTRIAL FIRMS[b]	COLLEGES AND UNIVERSITIES[c]	TOTAL[d]
	(Millions of Dollars)			
Department of Agriculture	449	1	209	667
Department of Commerce	203	37	44	319
Department of Defense	3,584	8,402	430	13,002
Department of Energy	352	1,718	255	4,689
Department of Health, Education and Welfare	769	171	1,970	3,665
Department of the Interior	269	80	41	435
Department of Transportation	116	192	21	418
Environmental Protection Agency	162	107	52	400
National Aeronautics and Space Administration	1,223	2,857	146	4,392
National Science Foundation	94	25	599	819

Source: National Science Foundation, *Federal Funds for Research and Development Fiscal Years 1978, 1979, and 1980* (Washington, D.C.: Government Printing Office, 1979).
[a] Estimates made by the National Science Foundation.
[b] In addition, $1,255 million was spent in federally financed research and development centers administered by firms.
[c] In addition, $1,411 million was spent in federally financed research and development centers administered by universities and colleges.
[d] Individual items do not sum to totals because some R and D is performed by other nonprofit organizations, state and local governments, and federally financed research and development centers.

leum, drug, rubber, primary metals, and food industries, among others, the percentage of R and D performance that is federally financed is much smaller. Thus, just as federally financed R and D is concentrated in a few areas, so federally financed R and D tends to be concentrated in a relatively few industries.

Turning from industry to the universities, our nation's colleges and universities are heavily dependent upon the federal government for R and D funds. About 70 percent of the R and D carried out by the colleges and universities is financed by the federal government. The leading source of these funds has been the Department of Health, Education, and Welfare (much of which is now included in the Department of Health and Human Services). Table 9.5 shows the forty universities that received the most federal obligations for R and D in 1978, and the amount each received. As would be expected, the leading research-oriented universities, such as MIT, Harvard, Cornell, Michigan, and Stanford, tend to rank among the highest. In 1978, the one hundred universities and colleges at the top of this list received about 84 percent of the total federal

TABLE 9.4

Funds for R and D Performance,
by Industry and Source, 1978

INDUSTRY	INDUSTRY-FINANCED	FEDERALLY FINANCED	TOTAL
	(Millions of Dollars)		
Food and kindred products	a	a	431
Textiles and apparel	a	a	87
Lumber, wood products, and furniture	136	0	136
Paper and allied products	a	a	394
Chemicals and allied products	3,233	361	3,594
Petroleum refining and extraction	952	119	1,071
Rubber products	a	a	504
Stone, clay, and glass products	a	a	331
Primary metals	518	28	546
Fabricated metal products	361	36	397
Machinery	3,876	583	4,459
Electrical equipment and communication	3,767	2,976	6,743
Motor vehicles	3,334	449	3,783
Other transportation equipment	a	a	131
Aircraft and missiles	1,859	5,821	7,680
Professional and scientific instruments	1,515	174	1,689
Other manufacturing	271	9	280
Nonmanufacturing	605	547	1,152
Totals	22,105	11,301	33,406

Source: National Science Foundation, "Greatest Increase in 1978 Industrial R and D Expenditures Provided by 14% Rise in Companies' Own Funds," *Science Resources Studies Highlights,* February 20, 1980.
[a] Not separately available but included in totals.

obligations to colleges and universities. Since the mid-1960s, there has been some pressure to allocate such funds more evenly.

3. Federal R and D Contracts and Grants: Rationale and Incentives for Efficiency

Given that federal R and D contracts and grants to the private sector amount to about $20 billion per year, it obviously is important that we consider the reasons why support of this kind is in the public interest. The rationale for such support varies from one area of support to another. Many of the areas characterized by relatively large amounts of federally financed R and D are intended to provide new or improved technology for public-sector functions. National security and space

TABLE 9.5

*Total Federal Obligations for R and D
to the Forty Universities and Colleges Receiving
the Largest Amounts, Fiscal 1978*[a]

RANK AND UNIVERSITY	MILLIONS OF DOLLARS	RANK AND UNIVERSITY	MILLIONS OF DOLLARS
1 Johns Hopkins	196	21 Washington (St. Louis)	42
2 MIT	114	22 New York	38
3 Stanford	80	23 Texas	37
4 California, San Diego	77	24 Duke	35
5 Washington	76	25 Ohio State	34
6 California, Los Angeles	70	26 Rochester	34
7 Harvard	69	27 Colorado	32
8 Wisconsin	65	28 Purdue	30
9 Columbia	65	29 Utah	29
10 Minnesota	60	30 Yeshiva	29
11 Michigan	58	31 Caltech	28
12 Cornell	58	32 California, Davis	28
13 Pennsylvania	56	33 Northwestern	27
14 Yale	54	34 North Carolina	27
15 California, Berkeley	50	35 Baylor	26
16 Chicago	49	36 Pittsburgh	26
17 California, San Francisco	46	37 Arizona	26
18 Illinois	43	38 Miami	26
19 Southern California	42	39 Case Western Reserve	25
20 Penn State	42	40 Iowa	24

Source: National Science Foundation, "Federal Obligations to Universities and Colleges Continued Real Growth in FY 1978," *Science Resources Studies Highlights,* February 20, 1980.
[a]Of course, not all of these universities and colleges are in the private sector. About half are state universities.

exploration, for example, are public goods—goods where it is inefficient (and often impossible) to deny their benefits to a citizen who is unwilling to pay the price. For such goods, the government is the sole or principal purchaser of the equipment used to produce them; and since it has the primary responsibility for their production, it must also take primary responsibility for the promotion of technological change in relevant areas. Even though much of the R and D of this type is performed by the private sector, it is important to note that the primary objective of this R and D is not to promote technological change in the private sector but in the public sector. Although there is unquestionably some beneficial spillover, the benefits to the private sector seem decidedly less than if the funds were spent directly on private-sector problems.[1]

1. See Mansfield, *The Economics of Technological Change, op. cit.* (cited in Chapter 1, footnote 16); and Mathematica, "Quantifying the Benefits to the National Economy from Secondary Applications of NASA Technology," unpublished, 1975.

In other cases, the rationale for large federally financed R and D expenditures is some form of market failure. In the case of energy, for example, it has been claimed that the social returns from energy R and D exceed the private returns because of the difficulties faced by a firm in appropriating the social benefits from its R and D. Also, it has been argued that risk aversion on the part of firms may lead to an underinvestment (from society's point of view) in R and D. Further, the availability of energy is frequently linked to our national security.[2] In the case of agriculture, the fact that farms are relatively small productive units has been used to justify federally financed R and D. The argument that farms are too small to engage in an efficient R and D effort certainly was more compelling when there were fewer and smaller industries supplying agriculture. But according to many experts, there still seem to be important aspects of farming that are not reflected in obvious markets for these suppliers.

Finally, as we saw in Table 9.2, some federally financed R and D is directed toward the general advance of science and technology. Such expenditures seem justified because the private sector will almost certainly invest less than is socially optimal in basic research. This is because the results of such research are unpredictable and usually of little direct value to the firm supporting the research, although potentially of great value to society as a whole. In other words, basic scientific information has many of the characteristics of a public good.[3]

We shall return to the question of the rationale for federal support of R and D in the private sector; but for now, we turn our attention to the incentives for efficiency and cost reduction in federally financed R and D. In a free-enterprise economy, there are important incentives for efficiency, one of the most important being that a firm can increase its profits (or reduce its losses) by reducing its costs. In other words, since firms under normal market conditions use fixed price contracts, increased efficiency means increased profit. Unfortunately, such incentives, which are so important in most areas of the economy, cannot be transferred at all easily to research and development, because R and D is so risky that fixed price contracts are generally not feasible. It is very difficult to establish a contract whereby the contractor agrees to obtain a certain quantum of information or to develop a certain product or process for a fixed price, because it is so difficult for the contractor to estimate how much it will cost to achieve this result. Thus, many government contracts for research and development are basically geared to reimburse the con-

2. For example, see John Tilton, *U. S. Energy R and D Policy* (Washington, D.C.: Resources for the Future, 1974).

3. Arrow, "Economic Welfare and the Allocation of Resources for Invention," *op. cit.* (cited in Chapter 7, footnote 1); and Nelson, "The Simple Economics of Basic Scientific Research," *op. cit.* (cited in Chapter 3, footnote 16).

tractor for whatever his costs turn out to be (within reason) to achieve the desired result. As is well known, these costs often tend to be much higher than are initially estimated. Alternatively, for some types of R and D, a certain contract amount is stipulated, and the contractor is expected to achieve as much as he can with that amount. In either case, the incentives for reducing costs undoubtedly are less than they would be if a fixed price contract of the ordinary sort were feasible.

However, this does not mean that there are no incentives for efficiency. In particular, if the award of new contracts is known to depend, at least in considerable part, on past performance, this can be a very important incentive. But for this incentive to operate, at least two conditions must be met. First, the contracting government agency must be in a position to judge the contractor's performance reasonably well. Clearly, this is not as easy as it may seem, since apparent failure may be due as much to luck as to lack of skill, and since the product of a research project may be difficult even for leading experts to evaluate. Second, there must be a reasonable amount of competition among potential contractors. If the government allows itself to get locked in to particular contractors, this incentive cannot operate at all well. Based on the studies at RAND,[4] by Peck and Scherer,[5] and by others, the problem of creating adequate incentives for efficiency in government-funded R and D carried out in the private sector is very real and very difficult to solve in anything other than a very approximate way. Certainly, however, the government should make sure that reasonably objective and unbiased judgments are made of contractor and grantee performance and that competition is encouraged wherever possible. Although these measures will not solve the problem, they will be a step in the right direction.[6]

4. Patents, Tax Incentives, and Other Existing Policy Instruments

Federal contracts and grants for R and D are by no means the only way in which the federal government currently supports R and D activities in the private sector. In this section, we provide a brief (and neces-

4. For example, see Burton Klein, "The Decision Making Problem in Development," *The Rate and Direction of Inventive Activity* (Princeton: Princeton University, 1962); and Thomas Marschak, Thomas Glennan, and Robert Summers, *Strategy for R and D* (New York: Springer Verlag, 1967).

5. M. J. Peck and F. M. Scherer, *The Weapons Acquisition Process* (Cambridge: Harvard University, 1962).

6. But the competition obviously should be real, not just a façade. The encouragement of many proposals that have no chance of being accepted to give the appearance of competition merely results in additional social waste. See Mansfield, Rapoport, Schnee,

sarily sketchy) description of some of the other important ways that the federal government provides such support.

The Patent System / The U.S. patent laws, which were discussed in some detail in Chapter 7, grant an inventor exclusive control over the use of his invention for seventeen years, in exchange for his making the invention public knowledge. Proponents of the patent system argue that these laws are an important incentive for invention, innovation, and early disclosure of new technology. Critics of the patent system stress the social costs arising from monopoly and question the importance of patents as an incentive in many parts of the modern economy. Few critics, however, would go so far as to say that the patent system does not encourage additional R and D in at least some parts of our economy.[7]

Tax Laws / The tax laws have provided some stimulus for private R and D. If the tax treatment of investment in plant and equipment and in R and D were neutral in terms of its effects on incentives, R and D would be classified as a capital investment, and depreciated over its useful life. Instead, our tax laws allow R and D expenditures to be treated as current expenses, which means that they are made more profitable relative to other forms of investment. Another provision of the Internal Revenue Code allows the sale of patents to be taxed at capital gains rates (which generally are lower than ordinary rates), even if the person is a professional inventor and in the business of making and selling patentable inventions.[8] (More will be said about recent changes in the tax laws in section 12.)

Regulation / Some aspects of federal regulation seem to have encouraged R and D activities in the private sector. For example, with regard to the airlines, it has frequently been concluded that attempts (prior to deregulation) to keep prices above the competitive equilibrium level resulted in a high rate, perhaps too high a rate, of technological change and innovation. Obviously, however, this is not true of all regulated industries. For example, in the railroad industry, it is frequently claimed that regulation has dampened research and innovation, e.g., in the case of the Big John covered hopper grain cars. Despite recent studies of the Averch-

Wagner, and Hamburger, *Research and Innovation in the Modern Corporation, op. cit.* (cited in Chapter 1, footnote 6).

7. See Jesse Markham, "Inventive Activity: Government Controls and the Legal Environment," in *The Rate and Direction of Inventive Activity* (Princeton: Princeton University, 1962); Scherer, *Industrial Market Structure and Economic Performance, op. cit.* (cited in Chapter 7, footnote 22); and many of the other references in Chapter 7.

8. See Murray Weidenbaum, "Government Spending and Innovation," unpublished, 1973; and Edwin Mansfield, "Tax Policy and Innovation," *Science* 215 (1982):1365–71.

Johnson effect, regulatory lag, and a variety of other relevant considerations, we know very little about the effects of various kinds of regulation on R and D in the private sector. This is unfortunate since about 10 percent of the nation's gross national product arises from the regulated industries, and since regulation has effects throughout the economy.

In addition to the older types of economic regulation, a variety of newer types of regulation, dealing with environmental protection, health, and safety, have been enacted. According to many industry groups, these regulations have resulted in disincentives for innovation because they have forced firms to devote technical and other resources to comply with the regulations, rather than to develop new products and processes. Also, these regulations sometimes increase the length of time until a new product can be marketed, and increase the uncertainties concerning the profitability of innovation. Unfortunately, although there is a widespread feeling that some regulations reduce the rate of innovation, no quantitative estimates of the effects of most regulations are available.[9]

Antitrust Laws / Our nation's antitrust policies seem to have important effects on research and innovation in the private sector. Although the evidence is limited, it appears that relatively strong competition tends to promote research and development, so long as firms are above some threshold size. Since it appears that new entrants are often significant sources of innovation, it seems important to eliminate unnecessary barriers to entry. However, the effects of antitrust policy are certainly not unmixed. For one thing, antitrust policies may cut the incentive of the dominant firm (or firms) in an industry to generate relatively rapid technical advance. Also, the fact that antitrust policy is at odds with the patent system may in some cases reduce the incentives for R and D in some industries.[10]

Technology Transfer Programs / The government currently invests in a number of activities to transfer the results of government R and D to the private sector. To the extent that these activities are effective, they are likely to encourage private R and D. Perhaps the best known of these activities is NASA's technology utilization program. This program has

9. See William Capron, ed., *Technological Change in the Regulated Industries* (Washington, D.C.: The Brookings Institution, 1971); Mary Mogee, "Federal Regulatory Practice and Technological Innovation," *Serving Social Objectives via Technological Innovation* (Washington, D.C.: National Science Foundation, 1973); Roger Noll, "Government Policy and Technological Innovation," unpublished, 1975; and George Eads, "Regulation and Technical Change: Some Largely Unexplored Influences," *American Economic Review* 70 (1980):50–54.

10. See Scherer, *op. cit.;* Markham, *op. cit.;* and Noll, *op. cit.*

included a number of research institutes and universities. For example, the Midwest Research Institute and the Aerospace Research Applications Center at Indiana University have received information concerning technological developments in the space program, and disseminated them to private industry. The success and effectiveness of this dissemination program, and others of a similar type, are difficult to measure.

Education / The federal government's policies to support education (in science and technology, and other fields as well) also encourage R and D in the private sector. The extent of private R and D is determined in part by the quantity and quality of scientific and engineering talent available in the society. Better-educated managers and workers seem to be better able to utilize research results, and more inclined to invest in R and D. The links between education, science, and technology are important, and the federal government's attempts to strengthen education certainly have helped to support R and D in the private sector.[11]

5. The Basic Economics of Government Support of Civilian Technology

In recent years, economists have made some attempt to determine, on the basis of general economic theory, whether it is likely that existing federal programs in support of civilian technology are adequate. In this section, we summarize some of the arguments bearing on this question. To begin with, it is generally agreed that, because it is often difficult for firms to appropriate the benefits that society receives from new technology, there may be a tendency for too few resources to be devoted to the development of new technology. It is also generally agreed that the extent to which these benefits are appropriable is probably related to the extent of competition faced by the potential innovator and to the kind of research or development activity in question. In particular, the more competition there is and the more basic the information, the less appropriable it is likely to be. However, this argument is blunted somewhat by the obvious fact that some inventive activity is carried on with little or no economic motive. Clearly, inventors and technologists are not motivated solely by dollars and cents.

Economists seem to agree that, because R and D is a relatively risky activity, there may be a tendency for firms to invest too little in it, given

11. See Edwin Mansfield, "Determinants of the Speed of Application of New Technology," in *Science and Technology in Economic Growth*, ed. B. Williams (London: Macmillan, 1973), and the literature cited there, as well as his "Education, R and D, and Productivity Growth," *op. cit.* (cited in Chapter 8, footnote 2).

that many firms seem to be averse to risk and that there are only limited and imperfect ways to shift risk. On the one hand, if firms are big enough so that their R and D program is reasonably large compared to particular projects, uncertainty is likely to be handled more effectively. On the other hand, since the threat of competitive innovation is an important stimulus to make firms more willing to accept the uncertainties involved in R and D, there are obvious disadvantages in firms becoming too large relative to the total market. In any event, it seems to be generally agreed that the riskiness of R and D is likely to result in less R and D than may be socially optimal.

Still another reason why there may be an underinvestment in particular kinds of R and D is that they may be characterized by significant indivisibilities. In other words, they may be characterized by economies of scale that prevent small organizations from undertaking them efficiently. This argument seems much more applicable to development than to research. It is important to recognize that, while firms may have to be a certain minimum scale to do many kinds of R and D effectively, this scale may be a relatively small share of the market. Furthermore, it is important to recognize that small firms have been responsible for many important innovations, while many big firms have concentrated on more minor improvement innovations. Nonetheless, bearing these qualifications in mind, it is often argued that some industries are so fragmented, they cannot do the proper amount of R and D.[12]

While the preceding arguments have a considerable amount of force, they by no means prove that there is presently an underinvestment in civilian technology. For one thing, these arguments generally are based on the supposition that markets are perfectly competitive, whereas in fact many important markets are oligopolistic. In oligopolistic markets, many economists believe that firms often stress product improvement as a form of rivalry, rather than direct price competition. Because of tacit agreement among the firms, this may be the principal form of rivalry, with the result that more may be spent on research and development than is socially optimal. One industry in which this is sometimes claimed to be true is the ethical drug industry. This is not, however, a proposition that is easy to prove or disprove.

Despite the arguments listed above, another reason why there may be no underinvestment in various forms of civilian technology is that the government is already intervening in a large number of ways to support civilian technology. For example, as we saw in section 4, there are already some general tax incentives that encourage R and D. Beyond this, in particular industries like aircraft, there are a host of government influ-

12. For a discussion of the considerations involved in this and the previous two paragraphs, see Noll, *op. cit.*

ences promoting R and D and technological change. For example, the government has paid for R and D related to aircraft. It has increased the demand for new airplanes by providing subsidies to the airlines and by regulating the airlines (prior to the chain of events leading up to their deregulation) in such a way as to discourage price competition. Of course, the aircraft industry is hardly typical in this regard, but, as we have seen, there is considerable government support for R and D of various kinds in the private sector, and it is not obvious, on a priori grounds, that the government has not already offset whatever latent underinvestment in R and D that was present in particular parts of the economy.[13]

Going a step further, some economists have argued that, even in the absence of oligopoly or government intervention, a private-enterprise economy might not underinvest in R and D. For example, it has been pointed out that the inventor might be in a position to predict and thus speculate on price changes resulting from the release of his new technology. In principle at least, this might offset the fact that he could not appropriate all of the benefits directly. But it is important to recognize how difficult it is to foretell what price changes will be, particularly since there are many factors other than the technology to be considered.[14]

In sum, there are several important factors related to the inappropriability, uncertainty, and indivisibility of R and D that seem likely to push toward an underinvestment in R and D by the private sector. But these factors may be offset, partially or fully, by oligopolistic emphasis on nonprice competition, by existing government intervention, or by other considerations. Thus, on a priori grounds, it is impossible to say with any reasonable degree of certainty whether there is an underinvestment in R and D in particular parts of the private sector.

6. Measurement of Social Benefits from New Technology: Agriculture

Since we cannot rely solely on a priori theorizing to tell us whether there is an underinvestment in R and D in the private sector (and if so, where it is most severe), we must turn to the available empirical studies of the returns from R and D of various types. These results should provide some information concerning what society has received from various forms of R and D investment in the past. Of course, there is a variety of problems in measuring the social benefits from new technology. Any

13. See George Eads, "U.S. Government Support for Civilian Technology: Economic Theory vs. Political Practice," *Research Policy* 3 (1974):1–16.

14. Jack Hirshleifer, "Where Are We in the Theory of Information?" *American Economic Review* 63 (1973):31–39.

innovation, particularly a major one, has effects on many firms and industries, and it obviously is difficult to evaluate each one and sum them up properly. Nonetheless, economists have devised techniques that should provide at least rough estimates of the social rate of return from particular innovations, assuming that the innovations can be regarded as basically resource-saving in nature.

To estimate the social benefits from an innovation, economists have used a model of the following sort. If the innovation results in a shift downward in the supply curve for a product (such as from S_1 to S_2 in Figure 9.1), they have used the area under the product's demand curve (DD^1) between the two supply curves—that is, $ABCE$ in Figure 9.1—as a measure of the social benefit during the relevant time period from the innovation. If all other prices remain constant, this area equals the social

FIGURE 9.1

Measurement of Social Benefits
from Technological Innovation

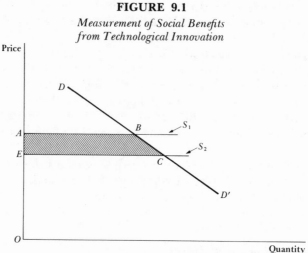

value of the additional quantity of the product plus the social value of the resources saved as a consequence of the innovation. Thus, if one compares the stream of R and D inputs (and other inputs) relating to the innovation with the stream of social benefits measured in this way, it is possible to estimate the social rate of return from the investment in the innovation.[15]

One of the first studies to use this approach was Griliches's study of

15. See E. Mishan, *Economics for Social Decisions* (New York: Praeger, 1972); and Edwin Mansfield, "Case Studies of the Measurement of Benefits from Scientific Information and Technological Innovation," presented at the First U.S.-USSR Symposium on the Economics of Information, Leningrad, 1975 (published in Russian).

hybrid corn.[16] Based on data concerning the increase in yields resulting from hybrid corn, the value of corn output each year, and the price elasticity of demand for corn, he could estimate the area corresponding to *ABCE* in Figure 9.1 each year. Then using data concerning the amount spent each year on hybrid-corn research, he could estimate the rate of return from the investment in hybrid-corn research, which turned out to be 37 percent. Clearly, a 37-percent rate of return is high. However, in evaluating this result, it is important to bear in mind that this is the rate of return from an investment which was known in advance to have been very successful. Thus, it is not surprising that it is high.

Another study, based on much the same principles, was carried out by Peterson[17] to estimate the rate of return from poultry research. This study, unlike the previous one, looked at the rate of return from all research in this particular area, successful or not. In other words, it included the failures with the successes. The resulting rate of return was 18 percent, which again is a rather high figure. However, as would be expected, this figure is lower than that for hybrid corn. A further study, by Schmitz and Seckler, used basically the same kind of techniques to estimate the social rate of return from the investment in R and D pertaining to the tomato harvester. The result depends on how long workers displaced by the tomato harvester remained unemployed, but the authors report that, even if the tomato workers received compensation of $2 to $4 million per year for lost jobs, the net social rate of return from the harvester would still have far exceeded 100 percent.[18]

It is important to recognize that all of the rates of return cited so far are average rates of return. That is, they are the average rate of return from all of the amounts spent on the relevant R and D. For many purposes, a more interesting measure is the marginal rate of return, which is the rate of return from an additional dollar spent. This is the measure that is most relevant in determining whether there is an underinvestment in civilian technology. If the marginal rate of return from investment in civilian technology is higher than the marginal rate of return from using the extra resources in other ways, more resources should be devoted to civilian technology. Thus, a very high marginal rate of return

16. Zvi Griliches, "Research Costs and Social Returns: Hybrid Corn and Related Innovations," *Journal of Political Economy* 66 (1958):419.

17. Willis Peterson, "Returns to Poultry Research in the United States," *Journal of Farm Economics* 49 (1967):656.

18. A. Schmitz and D. Seckler, "Mechanized Agriculture and Social Welfare: The Case of the Tomato Harvester," *American Journal of Agricultural Economics* 52 (1970):569–77. Since the concept of rate of return varies somewhat from study to study, the results are not always entirely comparable.

from investments in civilian technology is a signal of an underinvestment in civilian technology.

Using econometric techniques, a number of studies have estimated the marginal rate of return from agricultural R and D. One study, by Griliches,[19] investigated the relationship in various years between output per farm in a state and the amount of land, labor, fertilizer, and machinery per farm, as well as average education and expenditures on research and extension in a state. The results indicate that, holding other inputs constant, output was related in a statistically significant way to the amount spent on research and extension. Assuming a six-year lag between research input and its returns, these results indicate a marginal rate of return from agricultural R and D of 53 percent. Another study, by Evenson,[20] uses time-series data to estimate the marginal rate of return from agricultural R and D, the result being 57 percent. Also, Peterson's study of poultry R and D[21] indicates that the marginal rate of return for this type of agricultural R and D is about 50 percent. Schultz's study indicates a marginal rate of return of 42 percent.[22]

In sum, every study carried out to date seems to indicate that the average social rate of return from agricultural R and D tends to be very high. The marginal social rate of return from agricultural R and D also seems to be high, generally in the neighborhood of 40 to 50 percent. Of course, as stressed above, these studies are based on a number of simplifications, and it would be very risky to attach too much significance to them, since they are rough at best. All that can be said is that the available evidence, for what it may be worth, suggests that the rate of return from agricultural R and D has been high.[23]

7. Measurement of Social Benefits
from New Technology: Industry

Having summarized the available results concerning the social rate of return from R and D in agriculture, we must now provide the same

19. Zvi Griliches, "Research Expenditures, Education, and the Aggregate Agricultural Production Function," *American Economic Review* 54 (1964):961.

20. Robert Evenson, "The Contribution of Agricultural Research and Extension to Agricultural Production," Ph.D. diss., University of Chicago, 1968.

21. Willis Peterson, "The Returns to Investment in Agricultural Research in the United States," in *Resource Allocation in Agricultural Research* (Minneapolis: University of Minnesota, 1971).

22. Theodore Schultz, *The Economic Organization of Agriculture* (New York: McGraw-Hill, 1953).

23. Robert Evenson, Paul Waggoner, and Vernon Ruttan, "Economic Benefits from Research: An Example from Agriculture," *Science*, September 14, 1979, pp. 1101–7.

information for industry. The first study of this sort was made by Mansfield, Rapoport, Romeo, Wagner, and Beardsley,[24] and it included seventeen industrial innovations. These innovations occurred in a variety of industries, including primary metals, machine tools, industrial controls, construction, drilling, paper, thread, heating equipment, electronics, chemicals, and household cleaners. They occurred in firms of quite different sizes. Most of them are of average or routine importance, not major breakthroughs. Although the sample cannot be regarded as randomly chosen, there is no obvious indication that it is biased toward very profitable innovations (socially or privately) or relatively unprofitable ones.

To obtain social rates of return from the investments in each of these innovations, a model somewhat like that described in Figure 9.1 was used, except that the analysis was extended to include the pricing behavior of the innovator, the effects on displaced products, and the costs of uncommercialized R and D and of R and D done outside the innovating organization. The results indicate that the median social rate of return from the investment in these innovations was 56 percent, a very high figure. On the other hand, the median private rate of return was 25 percent. (In interpreting the latter figure, it is important to note that innovation is a risky activity.)

In addition, very rich and detailed data were obtained concerning the returns from the innovative activities (from 1960 to 1972) of one of the nation's largest firms. For each year, this firm has made a careful inventory of the technological innovations arising from its R and D and related activities, and it has made detailed estimates of the effect of each of these innovations on its profit stream. The average rate of return from this firm's total investment in innovative activities was computed for 1960–72, the result being 19 percent, which is not too different from the median private rate of return given in the previous paragraph. Also, lower bounds were computed for the social rate of return from the firm's investment, and they were found to be about double its private rate of return, which also agrees with the results in the previous paragraph.

To extend this sample and replicate our analysis, the National Science Foundation commissioned two studies, one by Robert R. Nathan Associates and one by Foster Associates. Their results, like those summarized above, indicate that the median social rate of return tends to be very high and much higher than the median private rate of return. Based on its sample of twenty innovations, Nathan Associates found the median social rate of return to be 70 percent and the median private rate of return to be 36 percent. Foster Associates, based on its sample of twenty

24. Mansfield, Rapoport, Romeo, Wagner, and Beardsley, "Social and Private Rates of Return from Industrial Innovations," *op. cit.* (cited in Chapter 7, footnote 31).

innovations, found the median social rate of return to be 99 percent and the median private rate of return to be 24 percent.[25]

The foregoing results pertain to the average rate of return. In earlier investigations based on econometric estimation of production functions, Mansfield[26] and Minasian[27] estimated the marginal rate of return from R and D in the chemical and petroleum industries. Mansfield's results indicated that the marginal rate of return was about 40 percent or more in the petroleum industry, and about 30 percent in the chemical industry if technical change was capital-embodied (but much less if it was disembodied). Minasian's results indicated about a 50 percent marginal rate of return on investment in R and D in the chemical industry.

In a more recent study, Terleckyj[28] has used econometric techniques to analyze the effects of R and D expenditures on productivity change in thirty-three manufacturing and nonmanufacturing industries during 1948–66. In manufacturing, the results seem to indicate about a 30 percent rate of return from an industry's R and D based only on the effects of an industry's R and D on its own productivity. In addition, his findings show a very substantial effect of an industry's R and D on productivity growth in other industries, resulting in a social rate of return greatly exceeding that of 30 percent. No evidence was found, however, demonstrating that government contract R and D has any effect on the productivity increase of the industries performing it.

Griliches[29] has carried out an econometric study, based on data for almost nine hundred firms, to estimate the rate of return from R and D in manufacturing. His results pertain only to the private, not the social, rate of return. He finds that the private rate of return is about 17 percent. It is much higher than this in chemicals and petroleum, and much lower than this in aircraft and electrical equipment. He finds that the returns from R and D seem to be lower in industries where much R and D is federally financed.

Based on computations for the economy as a whole, Denison concluded that the rate of return from R and D was about the same as the rate of return from investment in capital goods. His estimate of the

25. Foster Associates, *A Survey on Net Rates of Return on Innovations,* Report to the National Science Foundation, May 1978; and Robert R. Nathan Associates, *Net Rates of Return on Innovations,* Report to the National Science Foundation, July 1978.

26. Edwin Mansfield, "Rates of Return from Industrial Research and Development," *American Economic Review* 55 (1965):310.

27. Jora Minasian, "Research and Development, Production Functions, and Rates of Return," *American Economic Review* 59 (1969):80.

28. Terleckyj, *Effects of R and D on the Productivity Growth of Industries, op. cit.* (cited in Chapter 6, footnote 11).

29. Griliches, "Returns to Research and Development Expenditures in the Private Sector," *op. cit.* (cited in Chapter 6, footnote 11).

returns from R and D was lower than the estimates of other investigators, perhaps due to his assumptions regarding lags.[30] In his presidential address to the American Economic Association, Fellner[31] estimated the average social rate of return from technological progress activities, his conclusion being that it is "substantially in excess" of 13 or 18 percent, depending on the cost base, and that this is much higher than the marginal rate of return from physical investment at a more or less given level of knowledge.

To sum up, practically all of the studies carried out to date indicate that the average social rate of return from industrial R and D tends to be very high. Moreover, the marginal social rate of return also seems high, generally in the neighborhood of 30 to 50 percent. As in the case of agriculture, there is a variety of very important problems and limitations inherent in each of these studies. Certainly, they are very frail reeds on which to base policy conclusions. But recognizing this fact, it nonetheless is remarkable that so many independent studies based on so many types of data result in so consistent a set of conclusions.

8. Mechanisms of Government Support in Other Countries

Having discussed the available evidence bearing on whether or not there may be an underinvestment in civilian R and D of various kinds, we turn now to a brief description of some of the mechanisms used in three other countries—the United Kingdom, France, and Japan—to support R and D in the private sector.

United Kingdom / Like the United States, the United Kingdom has devoted a large share of its government R and D expenditures to defense and atomic energy (Table 9.6). At the same time, however, it has tried in a variety of ways to support civilian technology as well. The National Research and Development Corporation is a public corporation that supports the development of innovations by paying part or all of the development costs, licenses firms to exploit public sector innovations, and enters into joint ventures with private firms. The British government has provided financial support for small firms, research associations, and universities to further the practical applications of research. In 1970 it spent about $10 million to support research associations. In

30. Denison, *The Sources of Economic Growth in the United States, op. cit.* (cited in Chapter 1, footnote 20).

31. William Fellner, "Trends in the Activities Generating Technological Progress," *American Economic Review* 60 (1970):1.

addition, it has engaged in large programs of grants to industry for research on processes, provided "launching aid" for the development of civilian aircraft and engines, and lent advanced machine tools without fee to potential purchasers or users.[32]

Although it is difficult to evaluate programs of this sort, there seems to be a widespread feeling that Britain's programs have not been very successful. This is often attributed, at least in part, to the fact that the government has been too inclined to assume the entrepreneurial role and to engage in commercial development activities. The government has tended to commit itself to the full-scale development of particular technologies too soon and too massively. In other words, according to many experts in the United Kingdom and elsewhere, the British government has tended to engage in activities that might better have been left to the private sector.[33]

France / There have been a number of French programs to support civilian technology, particularly in high-technology fields or in fields thought to be important for industrial independence. There have been "thematic action programs" meant to coordinate applied work in interdisciplinary areas among several laboratories normally devoted to basic research. There have been "concerted actions," which established committees to support research in fields like molecular biology and applied mechanics. There has been an "aid to predevelopment" program, designed to help cooperative research organizations to development work on new technologies. There has been an "aid to development" program, providing loans (which may be forgiven) to cover development costs incurred by private firms.

Additionally, there has been a variety of tax incentives. All of the operating expenses in research and development have been fully deductible costs of doing business. Investments in buildings for R and D have been written off by 50 percent in the first year, the rest being depreciated over the structure's normal life. Firms that combined their R and D resources into a new organization have benefited from a tax deduction on their investment in the new organization. And to promote industrial funding of research institutions, there has been a 50-percent depreciation rate for shares taken in public or private R and D institutions, deductions of payments to R and D institutions from profits taxes (up to 3 percent of the firm's turnover), and exemption of taxes on legacies to approved R and D institutions.[34]

32. Herbert Hollomon *et al., National Support for Science and Technology: A Description of Foreign Experiences* (Cambridge: Center for Policy Alternatives, MIT, March 1974).
33. Robert Gilpin, *Technology, Economic Growth, and International Competitiveness* (Washington, D.C.: Joint Economic Committee of Congress, Government Printing Office, 1975).
34. Hollomon *et al., op. cit.*

TABLE 9.6

Percentage Distribution of Government R and D Expenditures among Selected National Objectives, by Country, 1976[a]

NATION	NATIONAL DEFENSE	SPACE	ENERGY PRODUCTION	ECONOMIC DEVELOPMENT	HEALTH	COMMUNITY SERVICES	ADVANCEMENT OF KNOWLEDGE[b]	TOTAL
United States	51	13	9	9	10	5	4	100
West Germany	12	5	11	13	3	5	51	100
France	30	5	9	23	4	2	26	100
Japan	2	5	8	23	3	3	55	100
United Kingdom	46	2	7	20	3	2	20	100

Source: Science Indicators, 1978 (Washington, D.C.: National Science Foundation, 1979).

[a]Figures pertain to 1974–75 for Japan, 1976–76 for the United Kingdom, and 1976–77 for the United States.

[b]Excludes general university funds for the United States.

In industries like electronics, French policy seems to have been to maintain at least one domestic supplier of each politically significant technology. In the eyes of many observers, this policy has had important drawbacks. According to Zysman:

> The dilemma has been that the protection and support required to produce specific products of interest to the state may, in fact, have weakened the firms that must be the long-term instruments of state policy. . . . Before the reality of technological independence, strong and innovative firms, can be realized, the symbol of particular goods produced by subsidized but feeble national companies may have to be abandoned.[35]

Japan / There has been a well-known Japanese emphasis on the importation of technology. The Japanese government has played a very important role in determining which technologies should be purchased from abroad and which firms should receive them. Besides relying heavily on foreign technology, Japan has spent significant amounts on R and D. As shown in Table 9.7, Japan's R and D expenditures, as a percent of

TABLE 9.7

*R and D Expenditures as a Percentage of
Gross National Product, 1974*

COUNTRY	TOTAL R AND D	NONMILITARY R AND D
United States	2.32	1.46
Japan	1.95	1.91
France	1.81	1.34
West Germany	2.26	2.27

Source: Science Indicators, 1978, op. cit.

gross national product, have been lower than in the United States or West Germany. But if one looks only at nonmilitary R and D, Japan's R and D expenditures, as a percent of gross national product, are higher than those of the United States. This, of course, is due to the fact that Japan spends very little on defense.

An interesting feature of Japan's technology policy is that a very low percentage of the nation's R and D has been financed by government. Japanese industry has supported a much larger share of the nation's R and D than has industry in the United States, the United Kingdom, or France. About three-fifths of the government's R and D expenditures

35. J. Zysman, "Between the Market and the State: Dilemmas of French Policy for the Electronics Industry," *Research Policy* 3 (1975):312–37. In 1981, the government of President Mitterrand said that it aimed at increasing R and D spending considerably, high-priority areas being biotechnology, electronics, robotics, and energy.

on economic development have been for the program of the Agency of Industrial Science and Technology, which has run about a dozen national R and D programs on electronic computers, electric cars, sea-water desalting, and other such topics. The projects have been chosen on the basis of their potential importance to the economy, and the appearance of market failure which has prevented the private sector from carrying them out. Also, the agency has provided subsidies (amounting to one-half of the costs) for particular development projects proposed by industry.

Japan also has used a variety of tax credits for industrial R and D. In 1967, it introduced a program whereby a firm was permitted a 25-percent tax credit for increases in R and D expenditures, the maximum tax credit generally being 10 percent of the corporate tax. Further, there has been accelerated depreciation for relevant plant and equipment of various kinds, and a partial tax exemption of receipts from foreign sale of technology.

Most observers seem to give high marks to Japan's programs in support of civilian technology. But it is difficult, particularly for outsiders, to characterize in a precise or detailed way the nature of some of these programs, since the Ministry of International Trade and Industry (MITI) has relied on informal guidance and intervention, as well as on formal controls, to influence the import of technology and the direction of civilian technology. However, one noteworthy feature of these programs is that they tended to view R and D as merely a part of the entire process of technological innovation, and that technological development has been viewed simultaneously with such other parts of the innovation process as investment, markets, and labor.[36] These views coincide with the emphasis in many recent studies of the innovation process.

9. Advantages and Disadvantages of Various Mechanisms for Federal Support

As stressed in sections 5 through 7, existing evidence is too weak to indicate with any degree of certainty whether there is an underinvestment in civilian technology of various sorts. All that can be said is that practically all of the studies carried out to date conclude that the average and marginal social rates of return from investments in innovation have tended to be very high. Nonetheless, most economists who have studied

36. M. J. Peck, "Infusion of Technology and the Mysteries of the Catch-up," unpublished, 1974; K. Oshima, "Research and Development and Economic Growth in Japan," in *Science and Technology in Economic Growth*, ed. B. Williams (London: Macmillan, 1973); and Gilpin, *op. cit.*

the question[37] seem to feel, on the basis of the existing evidence, that it is likely that some underinvestment of this sort exists. If so, it is important to consider the various means by which federal support for civilian technology might be increased. In this section, we discuss the major advantages and disadvantages of each of a number of mechanisms for federal support of private-sector R and D and innovation.

First, consider tax credits for privately financed R and D. Perhaps the most important advantages of this mechanism are that it involves less direct government control than some of the other techniques, and that in some respects it would be relatively easy to administer. Its most important disadvantages are that it would reward firms for doing R and D that they would have done anyhow, that it would not help firms that have no profits, and that it would be likely to encourage the same kind of R and D that is already being done (rather than the more radical and risky work where the shortfall, if it exists, is likely to be greatest). Furthermore, according to estimates made by former Secretary Peterson of the Department of Commerce, a 25-percent tax credit for R and D would mean that the Treasury would lose about $2 billion to $3 billion annually.[38] Also, any program of this sort might run into difficulties with regard to the definition of R and D, since firms would have an incentive to use as wide a definition as possible. Further, the costs of R and D are only part (and sometimes a very small part) of total innovation costs. More will be said about tax credits in sections 11 and 12. (And whereas only ordinary tax credits are considered here, *incremental* R and D tax credits are taken up there.)

Second, consider federal contracts and grants in support of civilian technology. This, of course, is the route taken by the Department of Defense and the National Aeronautics and Space Administration in much of their work. This is the route also taken by the National Research and Development Corporation in Britain and by some proposals in the United States.[39] It has the advantage of being direct and selective, but it can involve political problems in the choice of contractors, as well as problems relative to the disposition of patents resulting from such contracts and grants. As pointed out in Chapter 7, different government agencies have adopted different policies with respect to patents. Some, notably the Department of Defense, allow the title to the patent to remain with

37. See the papers by Fellner, Griliches, Mansfield, and Stewart in National Science Foundation, *Research and Development and Economic Growth/Productivity* (Washington, D.C.: Government Printing Office, 1972); R. Nelson, M. J. Peck, and E. Kalachek, *Technology, Economic Growth and Public Policy* (Washington, D.C.: The Brookings Institution, 1967); Arrow, *op. cit.;* and W. Capron, "Discussion," *American Economic Review* 56 (1966):508.
38. Weidenbaum, *op. cit.* This figure pertains to when Peterson was in office.
39. Nelson, Peck, and Kalachek, *op. cit.*

the contractor; others have retained title to the patents. There has been a long-standing argument over the relative merits of these different patent policies.[40] (As noted in Chapter 7, the 1980 changes in the patent laws try to make the policies of various agencies more consistent.) Another, more fundamental difficulty with this mechanism for supporting private-sector R and D is that it is so difficult to estimate the social costs and benefits of a proposed R and D project in advance. More will be said about this in section 11.

Third, the federal government could support additional civilian R and D by initiating and expanding work of the relevant sorts in government laboratories. This technique has the advantage of being direct and selective. But there are great problems in having R and D conducted by organizations that are not in close touch with the production and marketing of the product. It is very important that there be unimpeded flows of information and good coordination of R and D on the one hand, and production and marketing on the other. Otherwise, the R and D is likely to be misdirected, or, even if it is not, it may be neglected or resisted by potential users. This is a difficult enough problem for various divisions of a firm, and it would seem to be made worse if the R and D is done in government laboratories.[41]

Fourth, the federal government could insure a portion of private credit to firms for R and D and innovation costs. It is frequently claimed that the reluctance of lenders to extend credit to risky and long-term projects is an undesirable barrier to innovation. To the extent that this is the case, such a program might help to remedy the situation. The government could, for a fee, share the risk with the private lender for loans for R and D and related purposes. The advantages of such a program are that it would not commit the government to large expenditures, the administrative costs would be low, and there would be little federal interference in the lending decision. The disadvantages are that it results in a contingent liability for the Treasury, political problems could arise in awarding the loan insurance, and, most important of all, there is very little hard evidence that the capital markets operate so inefficiently (from a social point of view) that such a program is needed.[42]

Fifth, the federal government could use its own purchasing procedures to encourage technological change in the private sector. As shown in Table 9.8, the federal government's purchases of many kinds of goods and services are very substantial. The federal government could encour-

40. Mansfield, *op. cit.* (cited in footnote 1).

41. Pavitt, *The Conditions for Success in Technological Innovation, op. cit.* (cited in Chapter 1, footnote 12).

42. Rolf Piekarz, "Government Loan Insurance for Innovation," in *Serving Social Objectives via Technological Innovation* (Washington, D.C.: National Science Foundation, 1973).

TABLE 9.8

Government Sales as a Percent of Total Sales

PRODUCT LINE	PERCENT SOLD TO FEDERAL GOVERNMENT
Food and kindred products	1.86
Tobacco manufactures	3.53
Textile mill products	1.13
Lumber and wood products	0.96
Furniture and fixtures	1.99
Paper and allied products	0.82
Chemicals and allied products	1.53
Petroleum and coal products	1.45
Rubber and miscellaneous plastics products	2.57
Leather and leather goods	4.19
Stone, clay, and glass products	0.83
Primary metal industries	1.08
Fabricated metal products	3.20
Machinery except electrical	3.39
Electrical machinery and supplies	14.05
Transportation equipment	28.01
Instruments	11.05
Miscellaneous	1.97
Wholesale trade	1.60

Source: Study Group 13A on Commercial Products, *Final Report to the Commission on Government Procurement* (Washington, D.C.: February 1972), p. 42.

age innovation by using performance criteria, which specify the desired end result without limiting the design to existing products, rather than product specifications. Proponents of performance-based federal procurement argue that it will free industry to innovate (limited only by the requirement that it perform certain specified functions), encourage cost reduction for the government, and encourage the government to serve as a pilot customer for technical innovations in areas where it represents a big enough market or a market sufficiently free from local restrictions or codes to make it worth industry's while to innovate. The disadvantages of this mechanism are that performance criteria may be expensive to develop and administer, and that the procurement process may be made less efficient by adding innovation to the list of socioeconomic objectives that already influence this process.[43] Another suggestion is

43. D. Davenny, "Government Purchasing and Technological Innovation," in *Serving Social Objectives via Technological Innovation* (Washington, D.C.: National Science Foundation, 1973); and Weidenbaum, *op. cit.*

that the government could make greater use of life-cycle costs in purchasing decisions.

Sixth, the federal government could use its regulatory policies to try to encourage R and D in the private sector. According to some observers, some (but by no means all) of the federal regulatory agencies have, through their policies and procedures, tended to restrain or distort technological innovation in the industries they regulate.[44] Because relatively little is known about the effects of regulation on technological change, it is hard to specify exactly what changes might be effective (and cost-effective). Among the suggested alternatives are that technology advisers be located in the regulatory agencies, and that a technology impact statement be appended to all major regulatory decisions. More fundamentally, many groups have called for Congress to require regulatory agencies to pay more attention to the costs as well as the benefits of regulatory actions.[45]

Seventh, the federal government might establish prizes for important industrial innovations and developments. Such prizes would, of course, make privately financed R and D more attractive; if a firm or individual felt that a prospective R and D project might lead to results worthy of such a prize, the rewards would appear higher than without the prize. An important disadvantage of this mechanism is that it is so difficult to figure out which innovations are worthy of prizes and which are not. Given the problems in measuring the social importance of an innovation, this mechanism may not be as feasible as might appear at first glance.

10. The Domestic Policy Review on Industrial Innovation

In 1978 and 1979, the federal government carried out a Domestic Policy Review on Industrial Innovation. The Industry Advisory Subcommittee involved in this review prepared draft reports on (1) federal procurement, (2) direct support of R and D, (3) environmental, health, and safety regulations, (4) industry structure, (5) economic and trade policy, (6) patents, and (7) information policy. These drafts were discussed and criticized by the Academic and Public Interest Subcommittees involved in the review. Further, the Labor Subcommittee presented a report, as did each of a large number of government agencies. The overall result was a large and far-flung effort to come up with policy recommendations.

44. Committee for Economic Development, *Stimulating Technological Progress* (Washington, D.C.: Committee for Economic Development, 1980).

45. Mogee, *op. cit.;* Eads, *op. cit.* (cited in footnote 9); and Committee for Economic Development, *ibid.*

One theme that ran through the Industry Advisory Subcommittee's reports was that many aspects of environmental, health, and safety regulations deter innovation. As pointed out in previous sections, there is a strong feeling that this is the case in a number of industries, although it is recognized that we lack very dependable or precise estimates of the effects of particular regulatory rules on the rate of innovation. However, the recommendations of the Industry Advisory Subcommittee with respect to regulatory changes were met with considerable hostility by the Labor and Public Interest subcommittees. Another theme found in some of the Industry Advisory Subcommittee's reports was that tax credits for R and D expenditures should be considered seriously. Other groups, including the U.S. Treasury Department, did not warm to this proposal.

On October 31, 1979, President Carter put forth a number of proposals, based on the Domestic Policy Review. He asked Congress to establish a consistent policy with respect to patents arising from government R and D, and advocated exclusive licenses for firms that would commercialize inventions of this sort. Also, he asked the Justice Department to write guidelines indicating the conditions under which firms in the same industry can carry out joint research projects without running afoul of the antitrust laws. Further, to reduce regulatory uncertainties, he asked environmental, health, and safety agencies to formulate a five-year forecast of what rules they think will be adopted.

In addition, President Carter proposed the creation of four "generic technology centers" at universities or other sites in the private sector to develop and transfer technologies where the social returns far exceed the private returns. Each center would be jointly financed by industry and government. Also, he proposed a new unit at the National Technical Information Service to improve the transfer of technology from government laboratories to private industry, and he proposed that the National Science Foundation and several other agencies expand an existing program of grants to firms and universities that carry out collaborative research. He also asked that a program to support innovative small businesses (currently under way at the National Science Foundation) be expanded by $10 million in 1981. And he proposed that government procurement policies put more stress on performance standards rather than specific design specifications.[46]

46. Other proposals were made as well. See *The President's Industrial Innovation Initiatives*, Office of the White House Press Secretary, October 31, 1979.

11. General versus Selective Support Mechanisms

With the material in previous sections as background, we turn now to a discussion of some of the major considerations that should be kept in mind in appraising the policy options in this area. To begin with, it seems fair to say that most economists who have studied this problem have come away with the impression that our nation's programs in support of civilian technology are *ad hoc*, and that it is difficult to understand why we have allocated this support in the way that we have. For example, an enormous amount of support has been provided for civilian aviation technology, but very little has been provided for railroad technology; an enormous amount of support has been provided for agricultural technology, but very little has been provided for construction technology; and so on. (Perhaps this allocation of support can be defended, but I know of no serious attempts to do so.) Also, many economists who have written on this topic seem somewhat uncomfortable about the extent to which federal support of R and D in the private sector is related to a relatively few high technology areas. When one looks at federal expenditures for R and D performed in the private sector, the data, shown in Table 9.4, indicate that the lion's share goes to industries like aircraft, electrical equipment, and instruments. Yet the marginal rate of return from R and D may be higher in less exotic areas like textiles or machine tools than in these high technology fields.

If these misgivings are close to correct, it is likely that a general tax credit for R and D would be a relatively inefficient way of increasing federal support for R and D in the private sector. This is because, as pointed out in section 9, it would reward many firms for doing what they would have done anyway, and it would be likely to encourage the same sorts of R and D that are already being done. A tax credit for *increases* in R and D spending (that is, an incremental R and D tax credit) is less objectionable on these grounds, but it too is frequently regarded as inefficient because it is not sufficiently selective. To get the most impact from a certain level of federal support, it seems to be generally agreed that a more selective technique would be desirable.

However, to utilize more selective techniques, some way must be found to determine where the social payoff from additional federal support is greatest (or at least relatively high). The way that most economists would approach this problem is to use some form of benefit-cost analysis to evaluate the payoff from additional federal support of various kinds of R and D. Unfortunately, although such methods are of some use, they are not able to provide very dependable guidance as to how additional federal support for civilian technology should be allocated, due in large part to the fact that the benefits and costs from various kinds of R and D are very hard to forecast. As the Department of Defense knows so well, it is difficult indeed to forecast R and D costs. And even major corporations have difficulty in using various forms of benefit-cost analysis for R and D project selection, even though they have a much easier benefit concept to estimate than most government agencies do.[47]

Thus, the choice between the general and more selective forms of support is not as simple as it may seem at first. And when one recognizes that the estimates constructed to guide the selective forms of support may be biased for parochial, selfish, or political reasons, the choice becomes even more difficult. As George Eads[48] has pointed out, the organizations and individuals that benefit from, or have a positive interest in, a certain R and D program may inflate the benefits estimate by claiming various "secondary" or "external" benefits that in fact are spurious or at least exaggerated. Given that it is so hard to estimate with reasonable accuracy the true social benefits of various R and D programs, the result could be a distortion of social priorities, if the estimates are taken seriously. And if they are not taken seriously, it would be difficult to prove them wrong.

Another consideration also bears on this choice. As noted in section 7, some studies have concluded that an industry's R and D expenditures have a significant effect on its rate of productivity increase, but that the amount of federally financed R and D performed by an industry seems to have little or no such effect. In part, this may be due to the possibility that output measures in industries like aircraft are not reliable measures of social value. But it may also be due to a difference in the effectiveness of federally financed and privately financed R and D. At present, there is no way to tell how much of the observed difference is due to the latter effect; but if it turns out to be substantial, this would seem to favor tax credits rather than increased federal contracts and grants.

Based on the above considerations, it may be best to adopt a blend of selective and general forms of support. While selective forms of support have obvious advantages (where they are all appropriate), there is no reason why they cannot be supplemented with more general forms of support. Tax credits for increases in R and D spending are less objectionable than a tax credit for R and D spending, although the problems regarding the definition of R and D (and thus the measurement of increases in R and D expenditures) remain. If adequate measures were

47. Beardsley and Mansfield, "A Note on the Accuracy of Industrial Forecasts of the Profitability of New Products and Processes," *op. cit.* (cited in Chapter 6, footnote 10).
48. Eads, *op. cit.* (cited in footnote 13).

available to guide more selective forms of support, perhaps they alone could do the job; but such measures are presently in their infancy.

12. *The Economic Recovery Tax Act of 1981*

In August 1981, the Congress passed a major tax bill with at least four provisions that are relevant here.[49] First, the law provided a 25 percent tax credit for R and D expenditures in excess of the average of R and D expenditures in a base period (generally the previous three taxable years). Expenditures qualifying for the new incremental R and D tax credit are "in house" expenditures for R and D wages, supplies, and the use of equipment, 65 percent of the amount paid for contract research, and 65 percent of corporate grants to universities and certain scientific research organizations for basic research. The credit applies to expenditures made after June 30, 1981 and before 1986.

Second, firms are allowed to depreciate assets much more quickly than in the past. For R and D equipment, the new recovery period is three years.

Third, the law provides for a two-year suspension of regulation 1.861-8(which we discussed briefly in Chapter 5). The Secretary of the Treasury is asked to conduct a study of the impact of regulation 1.861-8 on R and D activities in the United States and on the availability of the foreign tax credit. Within six months of the enactment of the law, the secretary must submit a report of the study's findings to the Congress.

Fourth, firms are allowed a larger tax deduction than in the past for contributions of newly manufactured equipment to universities for research. In the past, the amount of the deduction was equal to the amount of the taxpayer's basis in the property. Now it equals the taxpayer's basis plus 50 percent of any appreciation (but not to exceed twice the basis).

It seems likely that these measures will have some positive effect on the rate of innovation, but existing knowledge and techniques do not permit us to estimate their quantitative effects with much precision. One important question, which time alone will help to answer is: to what extent will firms react to the incremental R and D tax credit by merely re-defining activities as R and D? Until some systematic analysis is carried out, policy makers will not have as much information as they need regarding the effects of these tax measures. It is important that economic research be carried out on this topic.

49. For a discussion of the effects of various existing and proposed provisions of the tax code, see Edwin Mansfield, "Tax Policy and Innovation," *Science, op. cit.*.

13. Technological Change and Antitrust Policy

Finally, it is essential to point out that our general economic policies have a notable impact on R and D and technological change in the private sector. Like other economic variables, the rate of technological change is influenced by the general economic climate or environment, which in turn is influenced by our general economic policies. Thus, our policies regarding market structure, competition, unemployment, inflation, foreign trade, and a host of other economic matters are important in this regard. In this section of this chapter, we take up the effects of one aspect of our general economic policies, namely, our antitrust policies. More will be said about the effects of other aspects of our general economic policies in Chapter 10.

There has been a considerable amount written by economists concerning the effects of market structure and antitrust policy on the rate of technological change. Although we are far from having final or complete answers, the following generalizations seem warranted, based on the available evidence.

First, the role of the small firm is very important at the stage of invention and the initial, relatively inexpensive stages of R and D. Studies by Jewkes, Sawers, and Stillerman, Hamberg, Mueller, and others[50] indicate that small firms and independent inventors play a large, perhaps a disproportionately large, role in conceiving major new ideas and important inventions. Further, although full-scale development often requires more resources than small firms command, the investment required for development and innovation is seldom so great or so risky that only the largest firms in an industry can do the innovating or the developing. Our own studies of the drug, coal, petroleum, and steel industries indicate that, in all of these industries, the firms that carried out the most innovations, relative to their size, were not the biggest firms.[51] Only in the chemical industry does it appear that the largest firm has done the most innovating relative to its size.[52]

50. J. Jewkes, D. Sawers, and R. Stillerman, *The Sources of Invention*, rev. ed. (New York: Norton, 1970); D. Hamberg, "Invention in the Industrial Laboratory," *Journal of Political Economy* 71 (1963):95; W. Mueller, "The Origins of the Basic Inventions Underlying DuPont's Major Product and Process Innovations," in *The Rate and Direction of Inventive Activity* (Princeton: Princeton University, 1962); Scherer, *op. cit.;* and Morton Kamien and Nancy Schwartz, "Market Structure and Innovation: A Survey," *Journal of Economic Literature* 13 (1975):1–37.

51. Mansfield, *Industrial Research and Technological Innovation, op. cit.* (cited in Chapter 1, footnote 1); and Mansfield *et al., op. cit.* (cited in footnote 6).

52. Mansfield, Rapoport, Romeo, Villani, Wagner, and Husic, *The Production and Application of New Industrial Technology, op. cit.* (cited in Chapter 1, footnote 5).

The available evidence does not seem to indicate that giant firms devote more resources, relative to their size, to inventive and innovative activities than their somewhat smaller competitors. There seems to be a threshold effect. A firm has to be a certain size to spend much on R and D (as defined by the National Science Foundation), but beyond a certain point, increases in size no longer bring a proportionate increase in R and D expenditures.[53] As would be expected, the threshold varies from industry to industry, but it appears that increases in size beyond an employment level of about 5,000 employees generally do not result in more than proportional increases in innovation inputs or outputs. Moreover, there is some evidence that the biggest firms produce less inventive and innovative output, per dollar of R and D, than smaller firms.

Turning from size of firm to industrial concentration (which can be quite a different thing), most studies of the relationship between industrial concentration and the rate of technological change conclude that a slight amount of concentration may promote more rapid invention and innovation. For example, very splintered, fragmented industries like construction do not seem to be able to promote a rapid rate of technological advance. But beyond a moderate amount of concentration, further increases in concentration do not appear to be associated with more rapid rates of technological advance. Thus, the evidence does not seem to indicate that very great concentration must be permitted to promote rapid technological change and the rapid adoption of new technologies.[54]

Several other points should be noted. First, new firms and firms entering new markets play a very important role in the process of technological change. Existing firms can be surprisingly impervious to new ideas, and one way that their mistakes and inertia can be overcome in our economy is through the entry of new firms. Second, cases sometimes occur where industries contain such small firms or markets are so fragmented that technological change is hampered. In such cases, as we pointed out in section 3 (in connection with agriculture), it may be good public policy to supplement the R and D provided by the private sector. Third, it is generally agreed by economists that the ideal market structure from the point of view of promoting technological change is one characterized by a mixture of firm sizes. Complementarities or interdependencies exist among firms of various sizes. There is often a division of labor, smaller firms focusing on areas requiring sophistication and flexibility and catering to specialized needs, bigger firms focusing on areas requiring larger production, marketing, or technological resources.

To sum up, the available evidence does not indicate that we must permit very great concentration of American industry in order to achieve

53. Scherer, *op. cit.* An exception here is the chemical industry.
54. Scherer, *op. cit.*

rapid technological change and the rapid adoption of new techniques. Instead, it seems to suggest that public policy should try to eliminate unnecessary barriers to entry and to promote competition in American industry. At the same time, it is worth noting that the effects of the antitrust laws are not unmixed. For example, the antitrust laws may reduce the incentives of the dominant firm (or firms) in an industry to innovate.

14. Summary

The federal government finances about $20 billion of the research and development carried out by industry and the universities. Many of the areas characterized by relatively large amounts of federally financed R and D are intended to provide new or improved technology for public-sector functions, such as defense and space exploration. In other cases, the rationale for large federally financed R and D expenditures is some form of market failure. Besides its grants and contracts for R and D, the federal government supports R and D activities in the private sector through the patent system, the tax treatment of R and D, some aspects of its regulatory and antitrust policies, its technology transfer programs, and its support for education. Taken as a whole, the government's support for R and D in the private sector is very substantial.

Nonetheless, many economists suspect that the United States may be underinvesting in civilian technology. On a priori grounds, it is impossible to say with any degree of certainty whether there is an underinvestment in R and D in particular parts of the private sector. There are several important factors, related to the inappropriability, uncertainty, and indivisibility of R and D, that seem likely to push toward such an underinvestment. But these factors may be offset, partially or fully, by oligopolistic emphasis on nonprice competition, by existing government intervention, or by other considerations. Thus, economists have turned for guidance to empirical studies of the social rate of return from past investments in civilian technology.

Practically every study carried out to date seems to indicate that the average social rate of return from investments in both agricultural and industrial innovation tends to be very high. The marginal social rate of return from investments in both agricultural and industrial R and D also seems to be high, generally in the neighborhood of 30 to 50 percent. Of course, as stressed above, these studies are based on a number of simplifications, and it would be very risky to attach too much significance to them, since they are rough at best. All that can be said is that the available evidence, for what it may be worth, suggests that the social rate of return from investments in civilian technology has been high. And this has been interpreted as a signal of an underinvestment in civilian technology.

Other countries, such as the United Kingdom, France, and Japan, have used a variety of government mechanisms to support civilian technology. The results seem to have been mixed. In the United Kingdom, the government seems to have committed itself to the full-scale development of particular technologies too soon and too massively. In other words, the government has tended to engage in activities that might better have been left to the private sector. In France, government policy seems to have been to maintain at least one domestic supplier of each politically significant technology. In the eyes of many observers, this policy has had major drawbacks. In Japan, the government's programs in support of civilian technology have received high marks. But it is difficult, particularly for outsiders, to characterize in a precise or detailed way, the nature of some of these programs, since MITI has relied on informal guidance and intervention, as well as on formal controls, to influence the import of technology and the direction of civilian technology.

A number of mechanisms for government support of civilian technology have been proposed in the United States—R and D tax credits, federal contracts and grants in support of civilian technology, expanded work of the relevant sorts in government laboratories, government insurance of private credit to firms for R and D and innovation costs, the use of government purchasing procedures and regulatory policies to encourage technological change in the private sector, and the establishment of prizes for important industrial innovations and developments. All of these mechanisms have problems. General forms of support tend to be relatively inefficient. Selective forms of support run into difficulties because it is so hard to forecast the social payoff from additional federal support for R and D of particular kinds. It seems likely that some combination of selective and general forms of support would be better than either alone. In 1981, an incremental R and D tax credit was passed by Congress. Economists should monitor the effect of this tax change (and the others in the 1981 tax act) on R and D expenditures and other relevant variables.

10 CONCLUSIONS

1. Introduction

Our purpose in this book has been to present new findings concerning two broad sets of topics. First, we studied the rates of international transfer of technology, the kinds of technology transferred overseas, the benefits of such transfer to the recipients, the effects of international technology transfer on U.S. research and development, the channels of international technology transfer, the resource costs of transferring industrial technology, the time-cost tradeoffs involved, and the overseas R and D expenditures by U.S.-based firms. Second, we studied the effects of the composition of an industry's or firm's R and D expenditures on its rate of productivity increase, the size and determinants of imitation costs, the characteristics of the nation's engineering labor force, and the nature and adequacy of existing federal programs affecting civilian technology. In this final chapter, we bring together our findings and discuss some of their implications.

2. The Increasing Rate of International Technology Transfer

One of the fundamental processes that influence the economic performance of nations and firms is technology transfer. Economists have long recognized that the transfer of technology is at the heart of the process of economic growth, and that the progress of both developed and developing countries depends on the extent and efficiency of such transfer. In recent years, economists also have come to realize (or rediscover)[1] the importance of international technology transfer in affecting the size and patterns of world trade. Yet there has been surprisingly little systematic, empirical analysis of the international transfer of technology.

1. Arthur Bloomfield, "The Impact of Growth and Technology on Trade in Nineteenth Century Economic Thought," *History of Political Economy* 10 (1978):608–35.

To help fill this gap, we carried out a number of studies of international technology transfer. Our results suggest that technology is being transferred across national boundaries more rapidly than in the past. In Chapter 2, based on a sample of chemical, semiconductor, and pharmaceutical innovations, we found this to be true even when a variety of other relevant factors were held constant. In considerable part, this is due to the growing influence of multinational firms, many of which are heavily involved in transferring technology. In Chapter 2, we found that U.S.-based multinational firms are transferring their technology to their foreign subsidiaries much more quickly than in the past. In 1969–78, about 75 percent of the technologies in our sample that were transferred to subsidiaries in developed countries were less than five years old; in 1960–68, the proportion was about 27 percent.

However, for technologies transferred to subsidiaries in developing countries or for those transferred through channels other than subsidiaries, there is no evidence in our sample of a reduction in the age of the transferred technology. The fact that the technologies licensed to, or jointly exploited with, non-U.S. firms were no newer in 1969–78 than in 1960–68 is worth noting, since many observers worry that U.S. firms may have come to share in this way more and more of their newest technologies with foreigners. In this connection, it might also be pointed out that the mean age of the technologies transferred in this way was much higher (about thirteen years) than the mean age of the technologies transferred to subsidiaries in developed countries (about six years) or in developing countries (about ten years).

Nations that spend relatively large amounts on R and D (in the relevant industry) tend to be relatively quick to begin producing a new product, even if they are not the innovator. This finding is analogous to our finding in an earlier book[2] that firms that spend relatively large amounts on R and D tend to be quick adopters of new technology developed by others. Both for entire nations and individual firms, R and D provides a window opening on various parts of the environment, and it enables the nation or firm to evaluate external developments and react more quickly to them. In some economic models, R and D is viewed as an invention-producing or innovation-producing activity. While correct as far as it goes, this view misses much of the point of R and D., which also is aimed at a quick response to rivals and at clever modification, adaptation, and improvement of their results.

In many important industries, like pharmaceuticals, international technology transfer is being prompted by the fact that companies have been carrying on increasing shares of their R and D overseas. As we saw

2. Mansfield *et al.*, *The Production and Application of New Industrial Technology, op. cit.* (cited in Chapter 1, footnote 5).

in Chapter 5, over 10 percent of the R and D carried out in 1980 by the firms in our sample was performed outside the United States. In some industries like pharmaceuticals, this percentage is much larger. When compared with the total R and D expenditures in various host countries, the size of overseas R and D is perhaps even more striking. In the early 1970s, about one-half of the industrial R and D performed in Canada and about one-seventh of the industrial R and D performed in the United Kingdom and West Germany were done by U.S.-based firms.[3]

The IBM Corporation has been a prominent example of how a firm's R and D activities can be integrated on a worldwide basis. Until 1961, IBM used its overseas laboratories to support the local market. But finding it difficult to make optimal use of these laboratories when their mission was limited in this way, the firm decided, when it developed the 360 line of computers, to bring the European laboratories into the worldwide development program. In the case of the 360 line, which consisted of six basic computers, each laboratory, whether American or European, was given a specific mission. For example, the smaller machine came from Germany, and the medium-size machine was designed in England.[4] By the mid-1970s, Chapter 5 indicated that about one-half of the firms we studied felt that worldwide integration of overseas and domestic R and D had been achieved.

3. The Product Life Cycle and the Internationalization of Firms' Operations and of Technology

The traditional way of looking at the process of international technology transfer has been built around the concept of the product life cycle. According to the product life cycle, there is a fairly definite sequence in the relationship between technology and trade, whereby the United States tends to pioneer in the development of new products, enjoying for a time a virtual monopoly. After an innovation occurs, the innovator services foreign markets through exports, according to this

3. When first established, many overseas R and D laboratories provided technical support to the company's overseas manufacturing operations and to overseas customers. As they grew older, some were given the responsibility of developing new and improved products for specific foreign markets, these products not being based directly on technology transferred from the parent. Some also developed new products for worldwide distribution and production, or developed new technology for the American parent corporation. In general, as pointed out in Chapter 5, overseas laboratories tend to focus on development rather than research, on product and process improvements rather than on new products and processes, and on relatively short-term technically safe work.

4. Mansfield, "Technology and Technological Change," *op. cit.* (cited in Chapter 1, footnote 46).

model. As the technology matures and foreign markets develop, companies begin building plants overseas, and U.S. exports may be displaced by production of foreign subsidiaries. The concept of the product life cycle has had a great influence in recent decades because it has been able to explain the train of events in many industries.[5]

Our results suggest that the situation may be changing, and that the product life cycle may be less valid than in the past. In the bulk of the cases we studied in Chapter 3, the principal channel through which new technologies were exploited abroad during the first five years after their commercialization was foreign subsidiaries, not exports. As indicated in the previous section, we found that about 75 percent of the technologies transferred by U.S. firms to their subsidiaries in developed countries during 1969–78 were less than five years old.

Thus, the "export stage" of the product cycle has often been truncated and sometimes eliminated. Particularly for new products, firms frequently begin overseas production within one year of first U.S. introduction.[6] In some industries, such as pharmaceuticals, new products commonly are introduced by U.S.-based firms more quickly in foreign markets than in the United States (due in part to regulatory considerations). For new processes, on the other hand, the "export stage" continues to be important. According to Chapter 3, firms are more hesitant to send overseas their process technology than their product technology because they feel that the diffusion of process technology, once it goes abroad, is harder to control. In their view, it is much more difficult to determine whether foreign firms are illegally imitating a process than a product.

To a considerable extent, this change in the process of international technology transfer and trade reflects the fact that many U.S.-based (and foreign-based) firms have come to take a worldwide view of their operations. At this point, many of them have in place extensive overseas manufacturing facilities. As emphasized in the previous section, many also have substantial R and D activities located abroad. Given the existing worldwide network of facilities and people, firms are trying to optimize the operation of their overall operations. This may mean that some of the technology developed in the United States may find its *initial* appli-

5. See Raymond Vernon, "International Investment and International Trade in the Product Cycle," *Quarterly Journal of Economics* 80 (1966):190–207; Raymond Vernon, ed., *The Technology Factor in International Trade* (New York: National Bureau of Economic Research, 1970); and Louis Wells, ed., *The Product Life Cycle and International Trade* (Boston: Harvard Business School, 1972).

6. According to a recent survey, this was true for about 40 percent of the innovations in 1971–75. See Davidson and Harrigan, "Key Decisions in International Marketing: Introducing New Products Abroad," *op. cit.* (cited in Chapter 2, footnote 18).

cation in a British subsidiary, or that an innovation developed in its British subsidiary may find its *initial* application in the firm's Brazilian subsidiary, and so on.

Another reason why the product life cycle is less valid than it used to be is that technology is becoming increasingly internationalized. For example, in the pharmaceutical industry, often it no longer is true that a new drug is discovered, tested, and commercialized, all within a single country.[7] Instead, the discovery phase frequently involves collaboration among laboratories and researchers located in several different countries, even when they are within the same firm. And clinical testing generally becomes a multicountry project. Even in the later phases of drug development, such as dosage formulation, work often is done in more than one country. In contrast, the product life cycle seems to assume that innovations are carried out in a single country, generally the United States, and that the technology resides exclusively within that country for a considerable period after the innovation's initial commercial introduction.

4. Should the United States Block the Outflow of Civilian Technology?

Since about 1970, regulation of commercial international technology transfer has been a public-policy issue in the United States. Traditionally, the American government has followed a "hands off" policy except for national-security considerations. Some groups argue that U.S. multinational firms are contributing to the deterioration of America's competitive position by transferring their technology to their overseas subsidiaries and by licensing it to foreigners. Although specific proposals for blocking (or at least slowing) the outflow of U.S. civilian technology have tended to be vague, there has been substantial sentiment for legislation to control the outflow of technology for commercial as well as national-security reasons.

Our findings in Chapter 2 provide some of the first empirical evidence concerning the effect of technology transfer (to an overseas subsidiary) on how quickly the technology leaked out to non-U.S. producers. Although in most cases the technology transfer had little or no effect on how rapidly the technology leaked out or an imitator appeared, it did have a substantial effect in a nontrivial proportion of cases. Specifically, in about one-fourth of the cases, the technology transfer seemed to has-

7. See National Academy of Engineering, *Technology Transfer from Foreign Direct Investment in the United States* (Washington, D.C.: National Academy of Sciences, 1976).

ten foreigners' access to these technologies by at least two and a half years. And in about one-third of the cases, the technology transfer resulted in at least a two-and-a-half-year reduction in the length of time elapsing before a foreign competitor imitated the innovation.

However, one cannot conclude from this that the United States should block the outflow of civilian technology. Our results in Chapter 3 do not support the suggestion of some economists that firms base their R and D decisions solely on the basis of expected domestic returns. On the contrary, according to the firms in our sample, about 30 percent of the anticipated returns from their R and D projects, on the average, were expected to come from foreign sources. Based on expected domestic returns alone, these firms would spend about 20 percent less on R and D than at present. Of course, these results do not contradict the hypothesis that firms sometimes pay less attention to foreign markets than to those at home. But they do indicate that this tendency is not so strong that public policy can assume that decreased opportunities for international technology transfer would have little or no effect on U.S. R and D expenditures. On the contrary, if these firms could not transfer or utilize their technology overseas, U.S. R and D expenditures would be likely to fall by about 20 percent.

Even if the government stemmed only the technological outflow to overseas subsidiaries, the effects would be substantial. If foreign subsidiaries could not be used, licensing or exports or joint ventures might be used instead. But in many cases, these other channels do not seem to be very good substitutes for foreign subsidiaries, in the eyes of the firms. If they could not use foreign subsidiaries as a channel, the best available estimate is that they would reduce their R and D expenditures by 12 to 15 percent on the average.

Besides their effects on U.S. R and D expenditures, measures designed to stem the outflow of U.S. civilian technology have other important disadvantages. For one thing, the history of technology tells us how difficult it is to prevent technology from crossing national boundaries. Many countries, including France, England, and Germany, have tried to do this, without success in the long run. And even if the United States proved capable of stemming the technological outflow, there is a good chance that it would provoke retaliation by other countries. The United States does not have a monopoly on advanced technology; indeed, it does not even have a substantial lead in many areas, according to expert opinion. For these reasons, as well as those given above, the federal government has not attempted to block the outflow of civilian technology.[8]

8. For a discussion of the ways in which some of the results of Chapters 3 and 5 were used in connection with this decision, see *Policy Research and Analysis*, National Science Foundation Program Report 3, no. 8 (December 1979). Of course, attempts to block the outflow of technology (to the USSR or to the gas pipeline) for political reasons do not concern us here.

5. Effects on Other Countries of the Outflow of U.S. Technology

As pointed out previously, technology transfer lies at the heart of the process of economic development. Innovations are primarily responsible for many increases in output per capita. How rapidly innovations spread—and thus raise per capita output in countries other than the innovating nation—depends on the process of technology transfer. The multinational firm is, of course, a major agent in the process of international technology transfer, but its role is highly controversial. Many host countries, although eager for modern technology, are suspicious of the activities of multinational firms.

One of the most important unanswered questions concerning the transfer of technology via the multinational firm is: How big have been the economic benefits to countries outside the United States arising from technology transfer of this sort? Put somewhat more precisely, how much lower would total output outside the United States have been if technology transfer of this sort had not occurred, but if the relevant technology and goods were perhaps available from the United States or elsewhere? In Chapter 2, we provided what seems to be the first quantitative evidence bearing on this question. Based on this analysis, we concluded that total annual output of countries outside the United States would have been at least $35 billion, or at least 1 percent, less if technology transfer of this sort had not occurred. (Note that this estimate is a lower bound.)

It is important to recognize that technology transfer of this sort can have beneficial effects on both developed and developing countries. The United Kingdom is the country (outside North America) where direct investment by U.S.-based firms has been greatest. In Chapter 2, we obtained information concerning the extent to which the technological capabilities of seventy major British firms during 1968–77 were affected by technology transfers by U.S.-based firms to their overseas subsidiaries. Industries where subsidiaries of U.S. firms account for a large share of the market seemed to be affected most by such technology transfer, but although there is some apparent tendency for the impact to be greater in more R and D–intensive industries, this tendency (when other factors are held equal) often is not statistically significant.

6. The Costs of International Technology Transfer

Although it is being transferred in greater quantities and more quickly now than in the past, technology frequently is neither easy nor cheap to transfer. In the sample of projects we studied in Chapter 4, technology transfer costs averaged about 20 percent of the total cost of establishing an overseas plant. Accordingly, although many economists regard existing technology as something that can be made available to all at zero social cost, this simply is untrue. Furthermore, transfer costs vary considerably, especially according to the number of previous applications of the innovation and how well the innovation is understood by the parties involved. It is equally inappropriate, therefore, to make sweeping generalizations about the size of the costs involved. For instance, technology transfer in chemicals and petroleum refining displayed relatively low transfer costs, presumably because it is possible to embody sophisticated process technology in capital equipment, which in turn facilitates the transfer process.

Our analysis in Chapter 4 of the determinants of transfer costs provides some findings with implications for the economics of development. The success of the more experienced enterprises, indicated by lower transfer costs, points toward economic models that emphasize the accumulation of skills, rather than fixed assets or capital, in facilitating the technology transfer process. This seems consistent with the findings of economic historians, such as Nathan Rosenberg.[9]

Our results also provide some managerial implications for the multinational firm. Consider the costs associated with separating production from development. The especially high cost of transfer before first application favors the development location, at least for production of initial units. Transfer costs will be lowered once the first production run has been commenced, a finding consistent with the product-life-cycle model. However, interindustry differences are important, and the costs involved in separating first production from development did not prove to be an insurmountable transfer barrier for an important subset of the sample projects.

A second implication is that since transfer costs decline with each application of a given innovation, technology transfer is a decreasing cost activity. This can be advanced as an explanation for the specialization often exhibited by engineering firms in the design and installation of particular turnkey plants, a characteristic particularly noteworthy of the petrochemical industry.

A third set of managerial implications relates to the criteria that might be used for the selection of a joint venture or licensing partner to utilize the innovating firms' technology abroad. Although the evidence is limited, gigantic firms do not seem to have any appreciable advantage over firms of moderate size in absorbing technology at a relatively low trans-

9. See Nathan Rosenberg, "Economic Development and the Transfer of Technology: Some Historical Perspectives," *Technology and Culture* 11 (1970):550–75.

fer cost. Highly research-intensive firms do not seem to have more than a small cost advantage in absorbing technology over firms with much lower R and D expenditures. However, manufacturing experience is important, especially for transferring machinery technology. In addition, there is evidence that transfers to governments in centrally planned economies involve substantial extra costs, sometimes because of high documentation requirements and differences in managerial procedures.

The cost of designing, constructing, and starting up a manufacturing plant abroad based on U.S. technology depends on the time span between project commencement and project completion. There is a time-cost tradeoff. According to Chapter 5, the elasticity of cost with respect to time for the international transfer of manufacturing technology generally seems to exceed one, at least in our sample. The elasticity of cost with respect to time depends on the duration of the preliminary planning stage of the project (relative to other stages), the extent to which the relevant technology has been applied before, the size of the primary transfer agent, the size of the total project costs, and whether or not the foreign market can be satisfactorily supplied by exports.

7. Technology Transfer, Market Structure, and the Patent System

In some instances, technology is transferred from one organization to another with the help, or at least the consent, of the owner of the technology. In other instances, much or all of the technology is revealed by the innovation itself. In some fields, reverse engineering—which, crudely speaking, involves analyzing and tearing a product apart to see what it consists of and how it is made—is a well-developed art. Even if a new product is not subject to reverse engineering, it may be possible to "invent around" the patents on which it is based, if it is patented. Because much of the relevant technology frequently is transferred (more or less involuntarily) to potential imitators, the costs of imitating an innovation frequently are substantially lower than the cost of the innovation itself.

Imitation costs are a very important, if neglected, topic. If imitation costs are substantially below the cost to the innovator of developing the innovation, there may be little or no incentive for the innovator to carry out the innovation. Imitation costs also affect concentration, since an industry's concentration level will tend to be relatively low if its members' products and processes can be imitated cheaply. But despite the major roles played by imitation costs, little or nothing has been known about their size or determinants. Based on Chapter 7, it appears that the average ratio of imitation cost to innovation cost is about 0.65, and the average ratio of imitation time to innovation time is about 0.70. The ratio of imitation cost to innovation cost depends on the proportion of the product's innovation cost that goes for research, on whether or not the new product is a new drug where the FDA requires the imitator to do extensive testing, and on whether or not the product consists of a new use of an existing material where some firm other than the innovator has patents on the material.

From the point of view of industrial organization, it is interesting that the average ratio of imitation cost to innovation cost in an industry seems to be reasonably highly correlated with the industry's concentration level. There are a great many other factors that influence an industry's concentration level, such as economies of scale. If future research confirms this finding, the implication would seem to be that the nature of the technology transfer process (including transfers that are both voluntary and involuntary from the point of view of the innovator) within an industry can explain much more of the differences among industries in concentration levels than is generally recognized.

Our findings also shed new light on the effects of the patent system, which (despite the system's age) are very imperfectly understood. Contrary to the assumption of some economic models, patented innovations are often imitated within a few years of their initial introduction. It is a myth that a patent nearly always results in a seventeen-year monopoly over the relevant innovation. Nonetheless, patents do tend to increase imitation costs, particularly in the pharmaceutical industry. In the sample as a whole, the median estimated increase (due to patents) in the ratio of imitation cost to innovation cost was 11 percent. According to the relevant firms, about one-half of the patented innovations in our sample would not have been introduced without patent protection. The bulk of these innovations occurred in the drug industry. Excluding drug innovations, the lack of patent protection would have affected less than one-fourth of the innovations in our sample, according to the firms themselves.

In addition, our findings should help to promote the search for more realistic and useful models of the innovative process. In recent years, there has been a tendency for some such models to assume that the innovator receives all of the benefits from an innovation and that imitation can be ignored. And in studies of optimal patent life, it is often assumed that the patent holder is free from imitation for the life of the patent. Although we understand how convenient such assumptions may be, our results suggest how considerably they depart from reality. (Recall, for example, that about 60 percent of the patented innovations in our sample were imitated within four years.) We hope that the excellent theorists who are currently working in this area will soon be able to relax these assumptions.[10]

8. Has There Been a Slowdown in the U.S. Rate of Innovation?

Long a technological leader among nations, America tended to take its technological preeminence for granted during the postwar period. No longer is this true. In the past decade, the United States has begun to look with increasing concern at its technological position. Large studies have been carried out by government agencies, business groups, and scientific and engineering organizations. Yet the truth is that we know very little concerning the state, determinants, and effects of technological change in the United States. And this is one of the important reasons why, despite these studies, it is so difficult to formulate an effective technology policy.

Much of the recent discussion in Washington and elsewhere has been based on the assumption that a slowdown has occurred in the U.S. rate of innovation. It is important to recognize that the evidence supporting this view is limited in both quantity and quality. Without question, the U.S. rate of productivity increase has fallen sharply in recent years. But as we saw in Chapter 6, a variety of factors (including declines in the rate of increase of the capital-labor ratio, the huge increase in oil prices, and the impact of various kinds of regulations) could have been partly responsible. Certainly the fact that there has been a productivity slowdown does not prove by itself that there has been a slowdown in the rate of innovation.

The patent rate in the United States has been falling since about 1969. As we saw in Chapter 7, in practically all of the fifty-two product fields for which data are available, the number of patents granted annually (by year of application) to U.S. inventors declined during the 1970s. However, the crudeness of patent statistics as a measure of the rate of innovation should be emphasized. The average importance of

10. For example, see P. Dasgupta and J. Stiglitz, "Industrial Structure and the Nature of Innovative Activity," *Economic Journal* 90 (1980):266–93; and G. Loury, "Market Structure and Innovation," *Quarterly Journal of Economics* 93 (1979):395–409. It should be noted that the innovator may receive far less than other firms that subsequently introduce the innovation in a somewhat different way or at a more propitious time. Even if the innovator is very successful, imitators may gain a substantial share of the benefits. Models that ignore considerations of this sort should be viewed with caution, since (among other reasons) they omit factors that are important impediments to firms' appropriation of the benefits from their innovations. By doing so, they may exaggerate the incentive for firms to invest in R and D and other innovative activities.

the patents granted at one time and place may differ from those granted at another time and place. The proportion of total inventions that are patented may vary significantly. Further, the decline in the patent rate seems to have occurred too in other countries, such as Germany, France, the United Kingdom, and Canada.

In some industries where one can measure the number of major innovations that are carried out per unit of time, there seems to be direct evidence of a fall in the rate of innovation. For example, in the pharmaceutical industry, the number of new chemical entities introduced per year in the United States has declined relative to the 1950s and early 1960s.[11] This measure suffers from the fact that it is difficult to find suitable weights for different innovations. Also, this measure overlooks the small innovations which sometimes have a bigger cumulative effect than some of the more spectacular innovations. But nonetheless the results are of interest.

Thus, many of the available bits and scraps of data point to a slackening in the pace of innovation in the United States. But the data are so crude and incomplete that it would be foolish to put much weight on them. This does not mean that there is no reason to be concerned about civilian technology in the United States. On the contrary, for reasons discussed in Chapter 9, we think that the United States may well be underinvesting in civilian technology. And more to the point, we think that in some important industries, such as pharmaceuticals, there may very well have been a fall in the rate of innovation. But it would be a mistake, in our opinion, to regard the above statistics as reasonably reliable measures, even though we tend to agree that there may have been some reduction in the overall rate of innovation.

However, it is very important to distinguish between various sectors of the American economy. In pharmaceuticals and agricultural chemicals, there may very well have been a decrease in the rate of innovation, due in considerable part to increases in regulatory requirements.[12] But in other parts of the economy, such as microelectronics, the rate of innovation seems to be hale and hearty. Recent advances in microprocessors and microcomputers are regarded by experts as extremely important.[13] And since the new microelectronic technology can be applied in a variety

11. Henry Grabowski, *Drug Regulation and Innovation* (Washington, D.C.: American Enterprise Institute, 1976).

12. For some opinions and evidence on this score, see National Academy of Engineering, *The Impact of Regulation on Industrial Innovation* (Washington, D.C.: National Academy of Sciences, 1979).

13. For example, see Organization for Economic Cooperation and Development, *Special Session on Impact of Microelectronics on Productivity and Employment* (Paris: Organization for Economic Cooperation and Development, May 1980).

of tasks, it should have widespread effects on many areas of the economy. Without denying that a slackening in the rate of innovation may have occurred in some industries, we feel that there is little evidence of such a slackening in other important industries.

Also, it is important to distinguish between a reduction in the rate of technological change in the United States and a reduction in the American technological lead over other countries. According to expert opinion, the United States no longer has the commanding technological lead it enjoyed during the 1950s or 1960s. As countries like Japan and West Germany recovered completely from World War II and devoted more of their energies and resources to transferring, adapting, and extending technology, they narrowed the technological gap considerably, and in some areas surpassed the United States. But this does not mean that the U.S. rate of innovation necessarily declined. Even in microelectronics, where there is little evidence of a decline in the U.S. rate of innovation, the gap between the United States and Japan has been reduced considerably, because, according to many experts, the Japanese, starting from a much lower level, have advanced more rapidly than we have.[14]

9. Possible Reasons for a Slowdown in the Rate of Innovation

There are a number of reasons why a slowdown in the U.S. rate of innovation might have been expected in recent years. The nation's total R and D expenditures, when inflation is taken roughly into account, seemed to remain essentially constant from 1966 to 1977.[15] If one accepts the sort of model presented in Chapter 6, the rate of innovation is directly related to the rate of increase of the stock of R and D capital.[16] Thus, the reduction during the 1960s and 1970s in the rate of growth of the stock of R and D capital would be expected to result in a reduction in the rate of innovation. Some economists, such as John Kendrick, have estimated that this reduction in the rate of growth of the stock of R and D capital was responsible for about one-fourth of the deceleration in the

14. For example, see G. Gregory, "The March of the Japanese Micro," *New Scientist*, October 11, 1979, pp. 98–101.

15. In the official figures, the GNP deflator is used to deflate R and D expenditures. The limitations of this procedure are discussed in section 14.

16. More precisely, these models imply that the rate of increase of total factor productivity is directly related to the rate of increase of R and D capital. But the idea certainly is that R and D influences total factor productivity in considerable measure via innovation. Note too that "the stock of R and D capital" is not the capital stock used in R and D. Instead, it is the sum of depreciated R and D expenditures. See footnote 11, Chapter 6.

growth of real output per hour of labor between 1948–66 and 1973–78. However, Kendrick assumes that all R and D, both privately financed and government financed, should count equally in his stock of R and D capital. For reasons discussed elsewhere, we believe that, for these purposes, a dollar of government-financed R and D should not count as the equivalent of a dollar of privately financed R and D. Consequently, we think that he overestimates the extent to which the reduction in the rate of increase of the stock of R and D capital was responsible for the productivity slowdown.[17]

What caused the fall in the rate of increase of the stock of R and D capital? As pointed out in Chapter 6, the cutbacks in government-financed R and D expenditures were due in considerable measure to the winding down of the space program and to reduced defense programs. The reasons for the slowdown in the growth of industry's real R and D expenditures are less obvious, but the available evidence indicates that the private rate of return from investments in R and D was lower during the late 1960s and 1970s than during early periods. In Chapter 6, we found this to be the case, based on econometric estimates. In an earlier study, we found the same thing, based on a detailed study of the R and D portfolio of one of the nation's largest firms.[18] To some extent, this reduction in the private rate of return may have been due to diminishing returns as more and more money was invested in R and D. To some extent, it may also have been due to increased regulatory requirements and other factors.

Besides the fall in the rate of increase of the stock of R and D capital, there has been a noteworthy shift in the composition of industrial R and D expenditures. Based on our results in Chapter 6, it is evident that there has been a shift away from more basic, long-term, and risky R and D projects in many industries, such as aerospace, chemicals, and rubber. In metals, aerospace, drugs, glass, rubber, machinery, and chemicals, there was a reduction between 1967 and 1977 in the proportion of R and D expenditures aimed at entirely new products and processes, rather than at product and process improvements. As pointed out in Chapter 6, these changes would be expected to depress the rate of innovation. In this connection, it is interesting to note that, in electronics (where, as noted above, there is no clear-cut evidence of a slowdown in the rate of innovation), there seems to have been no reduction in the proportion of

17. See John Kendrick, "Why Productivity Growth Rates Change and Differ," and Edwin Mansfield, "Comments on Why Productivity Growth Rates Change and Differ," both of which were presented at the Conference on Economic Growth held at the Institute of World Economics, University of Kiel, in June 1980. Also, see M. Ishaq Nadiri, "Sectoral Productivity Slowdown," *American Economic Review* 70 (1980):349–52.

18. Mansfield, "Comment," *op. cit.* (cited in Chapter 6, footnote 10).

R and D expenditures devoted to long-term, relatively risky, and relatively ambitious projects. (See Table 6.6.)

What caused this change in the composition of R and D expenditures in many industries? As indicated in Chapter 6, several reasons are given by R and D executives. The reason most frequently given (particularly in the drug and chemical industries) is the increase in government regulations, which in their opinion reduced the profitability of relatively fundamental and risky projects. Another reason is that breakthroughs were more difficult to achieve than in the past, because the field has been more thoroughly worked over. Another reason is the inflation in the 1970s, which will be discussed further in sections 13 and 14 below. Still another reason is that many firms have changed their view of how R and D should be managed.[19] In the 1950s and early 1960s, firms frequently did not try to manage R and D in much detail. Subsequently, many firms began to emphasize control, formality in R and D project selection, and short-term effects on profit. This shift in emphasis has tended to reduce the proportion of R and D expenditures going for basic and risky projects. Although there is general agreement that a greater emphasis on detailed management was justified, many observers wonder whether the pendulum may have swung too far.[20]

10. R and D and Productivity

In the previous section, we pointed out that shifts in the composition of R and D expenditures may have been partly responsible for the productivity slowdown. Our results provide the first evidence that the composition of an industry's R and D expenditures, as well as their size, influences the industry's rate of productivity increase. For both analytical and policy purposes, the total R and D figures are hard to interpret because they include such a heterogeneous mixture of activities. Basic research and applied research are mixed up with development. Long-term projects are mixed up with short-term projects. Projects aimed at small product and process improvements are mixed up with projects aimed at major new processes and products. To answer many important analytical and policy questions, it is essential to disaggregate R and D.

19. See Nason, Steger, and Manners, *Support of Basic Research by Industry, op. cit.* (cited in Chapter 6, footnote 27).

20. Besides the slowdown in the rate of growth of R and D capital and the change in the composition of R and D expenditures, the age composition of industry's engineering labor force has changed dramatically, with possible negative effects on the rate of innovation. As pointed out in Chapter 8, relatively young engineers have declined considerably as a percent of the total, due to the slowdown in the rate of growth of engineering employment.

In Chapter 6, we found that there is a direct relationship between the amount of basic research carried out by an industry or firm and its rate of productivity increase when its expenditures on applied R and D are held constant. Whether the relevant distinction is between basic and applied research is by no means clear; basic research may be acting to some extent as a proxy for long-term R and D. Holding constant the amount spent on applied R and D and basic research, an industry's rate of productivity increase between 1948 and 1966 seemed to be directly related to the extent to which its R and D was long-term.

One reason why much of the interindustry or interfirm variation in rates of productivity increase cannot be explained by variation in R and D expenditure is that the composition of R and D expenditure matters. Another reason is that an industry's or firm's rate of productivity increase does not depend only on the amount spent on R and D in the United States. At present, international technology flows are completely ignored in practically all econometric models designed to relate R and D to productivity increase. The inadequacy of this assumption seems obvious. As shown in Chapter 5, U.S.-based firms carry out a significant amount of R and D overseas, and this R and D has an effect on the rate of productivity increase in the United States. Even more important is the effect of R and D carried out by one organization in one country on technological advance and productivity increase in another organization in another country. For example, productivity increase in the American chemical industry was certainly influenced by the work of Ziegler in Germany and of Natta in Italy.

Still another reason for the large amount of unexplained variation in recent years is the emergence of various new kinds of regulation, which have been adopted for such purposes as environmental and consumer protection and occupational health and safety. Some of these regulations reduce the effect of R and D on measured productivity growth because they dictate that firms divert their R and D resources to comply with regulatory programs. In some industries like drugs and agricultural chemicals, regulations have resulted in considerable increases in the amount of R and D resources (and time) required to develop a new product. According to many firms, the regulatory agencies also have discouraged innovation because of the uncertainty and instability of regulatory standards. It might be expected that these regulations would result in a poorer relationship between R and D and productivity increase. In fact, recent studies (including those in Chapter 6) indicate that the relationship is poorer.[21] The advent of these regulations is by no means the

21. According to Zvi Griliches, there may have been a reduction since the late 1960s in the productivity of R and D. The factors discussed in previous paragraphs in the text may help to explain such a change over time, although they may be only a small part of

only reason for the poorer relationship, but it may be a contributing factor.

11. Public Policy toward Civilian Technology: Some Suggested Guidelines

For well over a decade, a number of economists have warned that the United States may be underinvesting in civilian technology.[22] The reasons for this suspicion were discussed in detail in Chapter 9. Among other things, these economists point out that the marginal social rates of return from investments in civilian technology have been very high, both in agriculture and industry, according to practically every study that has been carried out. Of course, each of these studies has a number of limitations, but it nonetheless is remarkable that so many independent studies based on so many types of data result in so consistent a set of conclusions. The evidence on this point has increased very substantially in the past few years, but economists still are unable to specify with precision where the underinvestment, if it exists, is greatest. However, for reasons discussed above, it seems likely that the gap tends to be greatest for long-term, relatively basic projects.

There are a variety of ways that the government can stimulate additional R and D in the private sector—tax credits, R and D contracts and grants, expanded work in government laboratories, loan insurance for innovation, purchasing policies with greater emphasis on performance criteria and life cycle costing, altered regulatory policies, and prizes. As

---

the story. The changes over time in the composition of R and D expenditures may have made the interindustry differences in the stock of R and D capital less meaningful. Also, the increased importance of international technology transfer may have made the domestic stock of R and D capital less relevant. Because of these changes, the conventional figures concerning the stock of domestic R and D capital may, in effect, have contained larger measurement errors in recent years, with the result that the apparent productivity of R and D may have been biased toward zero. Also, the emergence of new kinds of regulation could help to explain such a change. But until much more work is done, there is no way to tell how much (if any) of this change is really due to these factors. In fact, as Griliches points out, the change itself may be due to "data difficulties and the confusion of the times." See Zvi Griliches, "R and D and the Productivity Slowdown," *American Economic Review* 70 (1980):343–48.

22. For example, see Edwin Mansfield, "Contribution of R and D to Economic Growth in the United States," *Science*, February 4, 1972, 487–94; his "Federal Support of R and D Activities in the Private Sector," in *Priorities and Efficiencies in Federal Research and Development*, Joint Economic Committee of Congress, October 29, 1976; and Nelson, Peck, and Kalachek, *Technology, Economic Growth, and Public Policy, op. cit.* (cited in Chapter 9, footnote 37). For a discussion of these warnings in the business press, see "The Silent Crisis in R and D," *Business Week*, March 8, 1976.

we saw in Chapter 9, each of these ways has its problems, as well as advantages. An important problem with a general tax credit is its inefficiency; an important advantage is that it involves less direct government controls. An important problem with more selective support mechanisms is that it is so difficult to estimate in advance the social benefits and costs of particular types of R and D projects. As we concluded in Chapter 9, a combination of selective and more general forms of support may be best.

Although many economists suspect that there may be an underinvestment in certain areas of civilian technology, there is at the same time some concern that the federal government, in trying to improve matters, could do more harm than good. In this regard, the following five guidelines may be of use. First, to the extent that a program of this sort is selective, it should be neither large-scale nor organized on a crash basis. Instead, it should be characterized by flexibility, small-scale probes, and parallel approaches. In view of the relatively small amount of information that is available and the great uncertainties involved, it should be organized, at least in part, to provide information concerning the returns from a larger program. On the basis of the information that results, a more informed judgment can be made concerning the desirability of increased or, for that matter, perhaps decreased amounts of support.

Second, any temptation to focus the program on economically beleaguered industries should be rejected. The fact that an industry is in trouble, or that it is declining, or that it has difficulty competing with foreign firms is, by itself, no justification for additional R and D. More R and D may not have much payoff there, or even if it does, the additional resources may have a bigger payoff somewhere else in the economy. It is important to recall the circumstances under which the government is justified in augmenting private R and D. Practically all economists would agree that such augmentation is justifiable if the private costs and benefits derived from R and D do not adequately reflect the social costs and benefits. But in many industries there is little or no evidence of a serious discrepancy of this sort between private and social costs and benefits. Indeed, some industries may spend too much, from society's point of view, on R and D.

Third, except in the most unusual circumstances, the government should avoid getting involved in the latter stages of development work. In general, this is an area where firms are far more adept than government agencies. As Keith Pavitt has put it, government programs in support of civilian technology "should be managed on an incremental, step-by-step basis, with the purpose of reducing key scientific and technical uncertainties to a degree that private firms can use the resulting knowledge to decide when (with their own money) they should move into full-scale commercial development."[23] Although there may be cases where development costs are so high that private industry cannot obtain the necessary resources, or where it is so important to our national security or well-being that a particular technology be developed that the government must step in, these cases do not arise very often. Instead, the available evidence seems to indicate that, when governments become involved in what is essentially commercial development, they are not very successful at it.[24]

Fourth, in any selective government program to increase support for civilian technology, it is vitally important that a proper coupling occur between technology and the market. Recent studies of industrial innovations point repeatedly to the key importance of this coupling. In choosing areas and projects for support, the government should be sensitive to market demand. To the extent that it is feasible, potential users of new technology should play a role in project selection. Information transfer and communication between the generators of new technology and the potential users of new technology are essential if new technology is to be successfully applied. As evidence of their importance, studies show that a sound coupling of technology and marketing is one of the characteristics that is most significant in distinguishing firms that are relatively successful innovators from those that are relatively unsuccessful innovators.[25]

Fifth, in formulating any such program, it is important to recognize the advantages of pluralism and decentralized decision making. If the experience of the last thirty years in defense R and D and elsewhere has taught us anything, it has taught us how difficult it is to plan technological development. Technological change, particularly of a major or radical sort, is marked by great uncertainty. It is difficult to predict which of a number of alternative projects will turn out best. Very important concepts and ideas come from unexpected sources. It would be a mistake for a program of this sort to rely too heavily on centralized plan-

---

23. Keith Pavitt, "A Survey of the Literature on Government Policy Towards Innovation," unpublished, 1975, p. 16.

24. See George Eads and Richard Nelson, "Government Support of Advanced Civilian Technology: Power Reactors and the Supersonic Transport," *Public Policy* (1971). Pavitt, *ibid.*, reports that, according to a study by Gardner, the British government since World War II has recovered less than one-tenth of its outlays on launching aid for aircraft and aircraft engines. For an interesting discussion, see B. Klein, *op. cit.*

25. See Christopher Freeman, "A Study of Success and Failure in Industrial Innovation," in *Science and Technology in Economic Growth*, ed. B. Williams (London: Macmillan, 1973); Mansfield *et al.*, *The Production and Application of New Industrial Technology, op. cit.*; and Mansfield, Rapoport, Schnee, Wagner, and Hamburger, *Research and Innovation in the Modern Corporation, op. cit.* (cited in Chapter 1, footnote 6).

ning. Moreover, it would be a mistake if the government attempted to carry out work that private industry can do better or more efficiently.

12. Importance of Investment in the Innovation and Diffusion Processes

The policy instruments discussed in the previous section are designed to influence research and development. Similarly, the 25-percent incremental R and D tax credit enacted in 1981, as well as most of the recommendations stemming from the Domestic Policy Review on Industrial Innovation, were directed at research and development. Without denigrating the role of R and D in the innovation process, it is important to recognize that it is only one stage of this process. Unless each of the other stages works properly, more R and D may do little good. If the resources devoted to other stages are expanded, this is likely to result in increases in R and D. In formulating policy regarding civilian technology, it is important that all stages of the innovation process be considered.

One of the most expensive parts of the innovation process is the construction of new plant and equipment. Detailed manufacturing drawings must be prepared, tooling must be provided, and manufacturing facilities often must be constructed. In a sample of thirty-eight innovations in the chemical, machinery, and electronics industries, this stage accounted for about 40 percent of the innovation costs, on the average.[26] Because investment in plant and equipment accounts for a relatively large proportion of the cost of many innovations, measures that encourage investment are likely to encourage innovation. And since the profitability of R and D is dependent upon the profitability of the entire business venture of which the R and D is part, measures that reduce the after-tax costs to the firm of plant and equipment are likely to increase the profitability of R and D.

Further, encouragement of investment in plant and equipment is likely to increase the rate of diffusion of new techniques, thus increasing the rate of growth of productivity. Many new techniques cannot be employed unless new plant or equipment is constructed and utilized. Consequently, public policies which encourage investment are likely to encourage both innovation and the diffusion of new technology.

In the United States, investment has been a relatively small percentage of total output. In manufacturing, whereas Germany has devoted

26. Mansfield et al., Research and Innovation in the Modern Corporation, ibid.

about 16 percent of its output to capital investment, and Japan has devoted about 29 percent, we have devoted only about 9 percent of our output to capital investment.[27] This is one of the reasons for our sluggish rate of productivity increase. Many observers believe that the United States should adopt measures to increase its rate of investment in plant and equipment.

In the early 1980s, there were many suggestions, particularly in the business community but also among academics, that some form of accelerated depreciation be adopted. According to its proponents, accelerated depreciation would increase the profitability of investment in plant and equipment, and increase the cash flow that firms could use for these purposes. In 1981, the Congress passed a major tax bill which allowed firms to depreciate assets much more quickly than in the past. Relatively little is known about the quantitative impacts of such tax changes on the rate of innovation, but one would expect them to have some positive effect on it. More research is needed to estimate these impacts.

13. Importance of Macroeconomic Policies

Our nation's technology policies cannot be separated from its economic policies. Measures that encourage economic growth, saving and investment, and price stability are likely to enhance our technological position. Just as many of our current technological problems can be traced to sources outside science and engineering, so these problems may be resolved in considerable part by improvements in the general economic climate in the United States. Indeed, improvements in our general economic climate may have more impact on the state of U.S. technology than many of the specific measures that have been proposed to stimulate technological change.

To illustrate how the general economic climate influences technological change, consider the effects of the high rates of inflation that the United States has experienced in recent years. Inflation that is high on the average tends to be very variable in its rate; and as Milton Friedman pointed out in his Nobel lecture,[28] this reduces the efficiency of the price system as a mechanism for coordinating economic activity. In particular, economists, both liberals and conservatives, worry about the effect of high rates of inflation on investment. Thus, Robert Nathan has stated that: "There are many serious consequences of an economic, social, and

27. Committee for Economic Development, *Stimulating Technological Progress, op. cit.* (cited in Chapter 9, footnote 44).
28. Milton Friedman, "Nobel Lecture: Inflation and Unemployment," *Journal of Political Economy* 85 (1977), pp. 466–67.

political nature flowing from high rates of inflation. Perhaps its most clearly identifiable negative impact has to do with investment. High interest rates, difficulties in floating equity securities, the tendency of government policies to fight inflation with recessions, the drop in the value of the dollar, all relate to inflation and all serve to discourage new investment."[29]

High rates of inflation, through their effects on investment and through other channels, also affect R and D. By themselves, research and development frequently are of little value to a firm. Only when they are combined with plant and equipment and with manufacturing, marketing, and financial capabilities do they result in a commercially meaningful new product or process. To the extent that inflation reduces investment rates, it tends to discourage the sort of R and D that requires new plant and equipment for its utilization. To the extent that inflation makes long-run prediction of prices and circumstances increasingly hazardous, it tends to discourage the sorts of R and D that are long-term and relatively ambitious. Indeed, many of the reasons why inflation adversely affects investment in plant and equipment hold equally well for investment in relatively ambitious R and D projects. This does not mean that firms necessarily cut back on their R and D expenditures in inflationary times. For example, according to NSF data, they increased their R and D expenditures by 14 percent between 1978 and 1979. But it does suggest that they often are less inclined to fund relatively ambitious R and D projects than would be the case under a regime of relative price stability. In the words of a General Electric executive, "the additional discounting now required to compensate for inflation is leading to even more emphasis on shorter-term programs where an adequate return can be projected."[30] Evidence that this effect has been widespread was presented in Chapter 6.

Besides affecting industry-financed R and D, inflation can have a negative impact on government-financed R and D. Faced with excessive inflation, governments may feel compelled to trim R and D budgets as part of an anti-inflationary fiscal policy. To the extent that this R and D would promote more rapid productivity increase in the longer run, this may have the unintended effect of lowering productivity growth and perhaps worsening inflation then.

To prevent misunderstanding, we must add that very high unemployment rates, as well as very high inflation rates, will tend to discourage innovation. When sales are depressed and the future looks grim, the

29. Robert R. Nathan, "Memoranda of Comment, Reservation, or Dissent," in Committee for Economic Development, *Stimulating Technological Progress, op. cit.*, p. 68.

30. Lowell Steele, *Hearings before the House Subcommittee on Science, Research, and Technology*, July 23, 1979, pp. 12–13.

climate for innovation is hardly bright. Neither severe and prolonged recession nor double-digit inflation constitute a benign climate for industrial innovation. We focus attention on inflation because its effects frequently tend to be somewhat more subtle than those of severe unemployment.

14. Inflation and R and D Statistics

During inflationary times, a particularly pernicious effect on R and D decision making in both the public and the private sectors arises because of the great difficulties in measuring the rate of inflation in R and D. In view of the inherent difficulties and the strong assumptions underlying the few alternative measures that have been proposed, the official government R and D statistics use the GNP deflator to deflate R and D expenditures. The relevant government agencies are well aware that the GNP deflator is only a rough approximation. For example, the Comptroller General's 1979 report on *Science Indicators* suggested the use of alternative price indices for R and D.[31] But little is known about the extent to which price indexes for R and D inputs, if they were constructed in various industries, would differ from the GNP deflator.

To help fill this gap, we carried out a small-scale effort, which is still under way. We obtained detailed information from over thirty firms in the chemical, electrical equipment, oil, primary metals, fabricated metal products, rubber, textiles, and stone, clay, and glass industries. These industries account for a large share of the privately financed R and D carried out by U.S. industry. The firms in our sample account for about one-ninth of all company-financed R and D in the United States. For each firm, a price index for R and D inputs (including scientists and engineers, support personnel, materials and supplies, the services of R and D plant and equipment and other inputs) was estimated. This was basically a Laspeyres index, 1969 being the base year and 1979 being the given year. Also, non-Laspeyres price indexes were estimated.

In practically all of the industries, the rate of increase of the price index for R and D inputs exceeded the rate of increase of the GNP deflator during 1969 to 1979. Only in the electrical equipment industry was the former less than the latter. Thus, for these industries as a whole, the official statistics concerning deflated R and D expenditures seem to overestimate the increase during this period in R and D performance, if these R and D price increases are reasonably accurate. Taking all of these industries together, deflated R and D expenditures increased (during

31. *Report by the Comptroller General on Science Indicators* (Washington, D.C.: General Accounting Office, September 25, 1979), pp. 46–47.

this ten-year period) by about seven percent based on the GNP deflator, but only by less than 1 percent based on these price indexes for R and D inputs. Take at face value, this seems to indicate that the bulk of the apparent increase in real R and D in these industries was due to the inadequacies of the GNP deflator. Of course, this result should be viewed with caution for a variety of reasons.[32] But it does illustrate how inflation can distort the basic statistics on which major policy makers depend.

15. Technology's Role in Revitalizing
the American Economy

The early 1980s have been a period of reexamination of the American economy. Beset by very high rates of inflation, high rates of unemployment, deficits in our balance of trade, basic structural problems in major industries like steel and autos, and very low rates of productivity increase, the American economy, long the engine on which our people (and to some extent other peoples as well) could count for material progress, has seemed to be in trouble. Whether this trouble will prove only temporary depends upon the policies adopted by our government, the actions taken by our firms, and the attitudes of our people. The policies, actions, and attitudes required to revitalize the American economy are of many types. Proper monetary and fiscal policies are essential. So is proper attention by firms to longer-range objectives and to such apparently mundane matters as quality control.

In reacting to our current problems, it is very important that policy makers, both in the public and private sectors, recognize the central role played by technology. As pointed out in earlier chapters, Edward Denison and others have estimated that 40 percent or more of the long-term increase in output per person employed in the United States has been due to technological change. Although estimates of this sort are subject to many limitations, it certainly is true that technological change has been responsible for a substantial share of our past economic growth. Moreover, the rate of innovation in the United States has an important influence on the competitiveness of U.S. goods in world markets and on our rate of inflation.[33] Thus, even though available measures of the rate of innovation are very crude at best, the fact that some of them seem to indicate a slowdown in the rate of innovation, at least in some important sectors of the economy, is cause for concern.

32. See Mansfield, "Research and Development, Productivity, and Inflation," *op. cit.* (cited in Chapter 1, footnote 46).

33. Edwin Mansfield, "R and D's Effects on Economic Performance," testimony before the Subcommittee on Science, Research and Technology, U.S. House of Representatives, June 18, 1980.

In formulating policies to stimulate the rate of innovation in some major industries, our nation's regulatory apparatus should be examined carefully. Firms have complained repeatedly about the inadequacies of existing regulatory policies and the distortions in the allocation of resources that have inadvertently resulted from these policies. Undoubtedly, some of these complaints are self-serving or groundless. But some appear to have substance. In principle, the regulatory agencies should view each regulation in terms of a benefit-cost analysis. That is, the social benefits of each regulation should be compared with the social costs, which include the adverse effects of the regulation on innovation. Of course, these costs and benefits will generally prove difficult to measure with accuracy, and there are lots of difficulties and pitfalls. But at least this is the right approach. Unfortunately, this approach has not always been followed in the past.[34]

It should also be noted that the reduction of the American technological lead presents opportunities as well as threats to the United States. The technological advances made by other countries can be of great value to the United States, both directly and because they stimulate us to adapt and extend their, and our, technology. This has been true in the past, and it will be true in the future. However, for these opportunities to be exploited fully, it is important that American firms be open, not hostile or indifferent, to foreign ideas. American firms may find it worthwhile to monitor more closely technological developments abroad. According to some observers, they are not as adept at this as their foreign competitors.[35]

There has been much talk in Washington and elsewhere concerning the measures that might be taken to reverse the apparent slowdown in the rate of innovation. Some proposals seem to view technology as largely exogenous to the economic system. Until only a few decades ago, economists themselves tended to view it this way, but they no longer do so. Particularly under recent conditions, such a view could result in very substantial mistakes. In our opinion, one of the most important things to stress is that this view is fundamentally incorrect. On the contrary, as emphasized in section 13, the U.S. rate of innovation is heavily dependent on the general economic climate in the United States.

Of course, this does not mean that many steps to encourage and support civilian technology directly would not be worthwhile. Although we know relatively little about the effects of many of these proposed

34. See George Eads, "Regulation and Technical Change: Some Largely Unexplored Influences," *op. cit.* (cited in Chapter 9, footnote 9).

35. For example, the Conference on U.S. Competitiveness at Harvard University in 1980 concluded that one cause of U.S. difficulties in this area is "our apparent inability to adequately scan and adopt foreign R and D."

measures, it seems likely that some of them are worth trying, subject to the general guidelines discussed above. Rather, the point is that, unless the United States encourages investment and economic growth and restores reasonable price stability, it seems unlikely that direct measures specifically designed to encourage and support civilian technology will have as much effect as is frequently assumed.

16. Limitations of Existing Knowledge and of Our Findings

In conclusion, we should add the customary warning that the studies contained in this volume are based on only a sample of industries—chemicals, petroleum, drugs, electronics, automobiles, soap, computers, fabricated metal products, machinery, textile, paper, rubber, instruments, electrical equipment, aerospace, metals, and stone, clay, and glass. Moreover, within each of these industries, the data pertain generally to only a sample of firms. Because the basic data are limited in this way and because the models used here are rough in many respects, our results must be regarded as tentative. Much more work is required before we attain a satisfactory understanding of technology transfer, productivity, and public policy toward civilian technology.

On the other hand, it is our belief that these studies have gone quite far beyond anything attempted heretofore in analyzing the relevant aspects of the process of technological change. Our findings are based on data obtained from literally hundreds of firms, many of whose engineers, scientists, and managers spent considerable amounts of time with us. We have obtained extremely detailed data concerning hundreds of R and D projects, as well as the decisions and activities involved in these projects. New types of data have been gathered, and new models have been proposed, to help answer a variety of questions, many of which were previously almost totally unexplored.

INDEX

adaptation process in technology transfer, 28–29, 47, 88
aerospace industry, 66, 126, 131, 221
  age and education of engineers in, 164–65, 166–67, 171
  engineering earnings in, 169–70
  forecasting errors in, 159–60, 171
Aerospace Research Applications Center, 183
AFL-CIO, 21
age distribution of engineers, 164–65, 171, 222n
Agency of Industrial Science and Technology, Japanese, 195
age of technology, 36–38, 48, 209, 211
  transfer costs and, 72–73, 85
agriculture, benefits from new technology in, 185–88, 206
Agriculture Department, U.S., 175
aircraft industry, 6, 114, 126–28, 175, 190
  federal support of R and D in, 184–85, 201, 202
American Economic Association, 191
Andrews, F., 164n
antitrust laws, 182, 200, 204–6
applied research, 4, 5, 28, 57, 105, 117, 121, 125n, 222–23
Armour, Henry, 124n
Arrow, Kenneth, 3, 65, 66, 132, 179n, 196n
Aspin-Welch test, 37n
average rate of return, 186–87, 189–90, 191, 206

Baranson, Jack, 15, 17n, 28n, 36, 62
basic research, 3–4, 28, 57, 222–23
  capital, 117–18, 130
  defined, 3–4, 108
  interfirm differences in productivity and, 123–25, 223
  interindustry differences in productivity and, 117–22, 223
  overseas vs. domestic, 105

productivity increase in manufacturing and, 108–31
Beardsley, George, 116, 160, 189, 202n
Behrman, Jack, 17n
Biemiller, A., 21n, 27n
Brash, D. T., 23n, 41n
Bureau of Economic Analysis, 123n
Bureau of Labor Statistics (BLS), 10, 110–11
  depreciation rates by, 123, 124n
  labor supply projections by, 157–58, 159, 171

Cain, G., 154n, 155n, 159n, 168n
Canada, 13, 14, 24, 140, 210, 219
  cost of R and D inputs in, 103–4, 107
  multinational corporations in, 22, 23
capacity transfer, 28–29, 47
capital, 10, 15, 109–10, 114, 220–21
  basic research, 117–18, 130
  physical, 9, 13, 117, 123n, 227, 228
capital-labor ratio, productivity growth and, 111, 129, 218
Carter, Jimmy, 200
Caves, Richard, 27n, 28n, 40, 41n, 55, 62n
  technology transfer to non-U.S. firms as viewed by, 45–46, 49
chemical industry, 18–19, 103, 114, 126–28, 129, 167, 190, 209, 221, 222, 223, 230–31
  age of engineers in, 164–65, 171
  expected returns from foreign markets in, 51–55
  federal funding of R and D in, 175–76
  forecasting errors in, 159, 160, 171
  interfirm differences in productivity in, 123–25, 130–31
  patents in, 135, 138
  technological innovations in, 6–7
  transfer costs in, 76–80, 81, 83, 84–86, 215
χ^2 tests, 37–38, 40, 106n
China, People's Republic of, technology transfer and, 74n